Series Editors

W. Hansmann
W. T. Hewitt
W. Purgathofer

W. Lefer and
M. Grave (eds.)

Visualization in
Scientific Computing '97

Proceedings of the Eurographics Workshop
in Boulogne-sur-Mer, France,
April 28–30, 1997

Eurographics

SpringerWienNewYork

Ass. Prof. Dr. Wilfrid Lefer
Laboratoire d'Informatique du Littoral, Université du Littoral,
Calais, France

Dr. Michel Grave
Office National d'Etudes et de Recherches Aerospatiales,
Châtillon, France

Typesetting: Camera ready by authors

Graphic design: Ecke Bonk

Printed on acid-free and chlorine-free bleached paper

SPIN: 10645470

With 92 partly coloured Figures

ISBN-13:978 - 3-211-83049-9   e - ISBN-13:978 - 3-7091-6876-9
DOI:10.1007/978 - 3-7091-6876-9

ISSN 0946-2767

Preface

The Eurographics working group on Visualization in Scientific Computing organized its eighth annual workshop in Boulogne-sur-Mer (France), on April 28-30, 1997. Considered as an emerging field when the first workshop took place in 90, Visualization is now a recognized research domain whose impact on human-data interaction is very promising. For seven years computer technology has considerably evolved both on hardware and software sides, allowing more and more sophisticated techniques to be designed and implemented. Of course many important issues remain. While a large part of today research is concerned with improving existing techniques and investigating new ones, we also deal with better integration of visualization tools with other computer technologies such as databases, networking or human-computer interfaces. Emerging fields such as Virtual Reality or the World Wide Web are also studied in the context of Visualization.

Contributions from 14 different countries were received, 22 of them were accepted for presentation during the workshop and a further selection reduced to 14 the number of papers finally kept for publication in this volume. The papers selected represent a real update to what has been already published and must be seen as presenting either usage of new technologies and development of new visualization techniques or enhancement of previously published ones. Papers cover many topics, mainly in three different categories:

- mathematical methods for extracting relevant visualizable information from scientific data sets are described, including new feature extraction mechanisms using wavelet representations or segmentation algorithms, as well as enhancements to popular algorithms like streamlines placement or particle path computation.
- some aspects of new computing technologies are approached in papers discussing visualization environments based on WWW tools or presenting advance user interfaces.
- finally some papers focus on applications of visualization in various domains such as finance, physics or space studies.

In addition, during the workshop, two panels were organised to discuss particular issues related to the use of new computing technologies for visualization. The first one was devoted to the use of the World Wide Web technology in visualization, with discussions on the client/server scheme and usage of techniques like Java Applets, ActiveX components or VRML viewers for enriching interaction functionalities of WWW based visualization tools. The second panel dealt with the use of Virtual Reality tools in the context of a visualization environment and participants exchanged valuable information on their experiences with CAVEs, Workbenches and other advanced devices. A short paper summarizing ideas investigated during these panels has been included in this book to keep track of these discussions and complement information provided during the workshop.

This volume covers a wide range of the areas of interest of the ViSC community today and provides researchers and engineers with valuable up to date information for setting up new powerful environments.

Many people contributed to the success of this workshop. Special thanks are due to the Visualization Group of the University of Littoral at Calais (F) who took care not only of the preparation of the workshop, but also of its progresses and of course of its conclusions, including gathering papers for this book. Mikael Jern and Jacques David should also be thanked for having brought to the attendees interesting information and having efficiently shared the panels.

Wilfrid Lefer and Michel Grave
Workshop co-chairs

Contents

WWW and Virtual Reality for Scientific Visualization

Jacques David and Michel Grave

CEA/DI Saclay & ONERA Châtillon, France

Abstract. Visualization in scientific computing integrates many different computer technologies. Among them two are evolving rapidly: the World Wide Web and Virtual Reality. These issues were discussed during paper presentations, and were even more precisely debated during dedicated panel sessions in the Eighth Eurographics Workshop on ViSC. This paper summarises and consolidates information that circulated among participants, but also provides additional sources of information on these topics.

1 World Wide Web

WWW Technology is improving and expanding quickly in all domains of Computing. Visualization in Scientific Computing is likely to get advantage of it, and this was analyzed during the Workshop. More precisely, three papers were presented and discussion took place during the panel session.
Many different aspects on WWW technology impact were addressed during the workshop, and are summarised here.

1.1 Client/Server architecture applied to ViSC platforms

The first clear advantage of applying WWW technology to VisC resides in the implementation of the Client/Server scheme, already known and applied successfully in many other different areas [16]. Like in others cases, applications are built with two main components:

- a Client (the WWW Browser), managing user's interfaces. It formats requests for a server and handles data returned, possibly by launching other programs (plug-ins) on the client site.
- a Server, answering to requests. It manages the processing tasks necessary to build answers to requests, and handles data return to the client.

A Visualization service reachable through WWW protocols can therefore in principle be used from any WWW browser, on any type of software and hardware platform conforming to WWW standards. This is for example discussed in the selected papers from the workshop [15][17], and in other references like [18]. Heterogeneity of hardware and software platforms can then be in large part hidden to the end users.

1.2   Protocols and distributed processing

The information exchange mechanisms between servers and clients are evolving quickly, enriching the potential functionalities of the applications set-up:

- the HTTP protocol and HTML language provide more and more advanced mechanisms for exchanging data (data download and upload, forms, ...) allowing many different data formats to be used.
- the CGI mechanism allows a client to execute programs on the server side, and to send parameters and data to these programs, in particular through HTML Forms.
- new extensions like scripts (JavaScripts, VBScripts), Java applets, Java Beans or ActiveX components handling give the possibility to download part of the processing and to actively manage user interactions on the client.

These features provide visualization applications designers a large choice for implementing distributed visualization systems over the WWW. The classical *Haber pipeline* for scientific visualization presented in [14] (or in a slightly modified version in [13]) can be cut at any point. This is the classical *fat client* or *fat server* paradigm well known for example in Databases applications [16].
The NPAC Visible Human Viewer [6] showed very early how at least the user interface of a browser in a large 3D data set could be implemented as a Java applet on the client side, and accessing the server only for getting data. Interactive graphical access to a Database for Microbial Genomes [5] or to molecular data [1] can also be found.
During the workshop, two different strategies were presented for distributing the work between a client and a server, and this was further discussed in the panel session. A first strategy [15][17][18], consists in executing all non graphical tasks on the server and to send to the client images or geometries (JPEG, VRML or other as discussed later). The client must be able to understand formats, and has therefore either to have the ad-hoc plug-ins, or to download the component to perform this task (Java Applet, ActiveX component, ...). This can be considered as a *visualization services* approach, where the server actually performs visualization tasks. A second strategy [15] consists in using the server as a software distribution source, where the client can get the modules it needs for performing itself the tasks. In this case again, this can be achieved through either dynamic download of components or pre-load of *plug-ins*.
Both strategies have their own advantages and drawbacks. In one case users all work on a single site, while in the other they download the necessary components from a single source. They therefore both greatly solve the problem of software update on users site, the only elements necessary to be really updated being the browser and a few plug-ins. However problems like overload of the servers and networks quickly appear.
The decision criteria include topics as different as localisation of data to visualize, computer loads, network bandwidth, access rights and security issues or licensing and copyrights rules.

1.3 Data formats

The nature of information returned by servers has a very high impact on the nature of interactions provided to the users. Each stage of the visualization pipeline [14][13] produces data of a specific type, like for example scientific data, 3D or 2D geometrical objects or images. If the pipeline is distributed between a server and a client, the closer data transmitted to the server will be to the top level (scientific data), the higher will be the interactivity capabilities on the client. For example if images are transmitted, rotations of a 3D object will have to be performed on the server, and therefore, response time to the user will probably be long. This is an other factor to take into consideration for choosing a distribution strategy.

The http protocol has been designed so that it can handle any kind of data types (with a MIME-like data *typing* mechanism). This means that any program on a client side can in principle be used as a browser extension (plug-in) for presenting data to the users.

The drawback of using too much this facility is that clients should have locally a wide range of programs already installed for being able to understand the data returned by servers.

Therefore, a limited subset of formats needs to be agreed on by a large community for limiting the number of plug-ins a client should have, or for limiting download of specific components.

In the WWW context, different levels of graphic formats are addressed, from images to 3D *active* geometries [11].

Image formats were not precisely adressed throughout the workshop. Concerning 2D graphics, Broadway [7] for example was only mentioned, probably because it was too new for having been used by the participants.

On the contrary, VRML received particular attention during the workshop, not only in already mentioned presentations [15][17], but also with an updated presentation of some earlier work [18], and in both panel sessions. Presentations corresponding to references [15] and [18] more precisely mentioned limitations of VRML 1.0 and the high potential of VRML 2.0 with the possibility to incorporate dynamics and script in it, as well with the data compression handling.

This capability to incorporate processing inside VRML files provides an other mechanisms for distributing work between a server and a client. VRML discussions showed clearly how borders were fuzzy between different application domains and for example provided a large bridge between WWW and Virtual reality.

2 Virtual reality

Virtual Reality is a new and embracing discipline, with soft delineations with respect to limits of the domain. The main characteristic which interests us here, is the whole rendering field, with or without immersive aspect. But rendering is also one strong topic for scientific visualisation, and even if the emphasis is not the same for the two disciplines - world-likeness, realism and immediate interactivity for VR, precision, fidelity, large data and exploration for ViSC- it appears rather obvious that both can collaborate and enrich each other.

Two papers on Virtual Reality were presented during the *VR and Visualization over the Web* session, and three related papers in other sessions. A fruitful informal exchange took place during the panel session in which light was focused onto this emerging domain. The following lines attempt to give a quick overview of the points that were evoked and point to complementary information.

2.1 Context

What is VR? The question arises from the ViSC users, as VR is a catch-all word, and before we can easily speak of it, terms and concepts must be defined. Classically, VR is characterized by *the three I*: Immersion, Interaction, Imagination. The operative concept is the notion of the virtual word, which must be made sensible to the *cyber* traveller, in such a way that he is, by way of non-intrusive peripherals (immersion) and his own enthusiasm (imagination), able to operate (interact) within this world as he would do in a real world.

Current representations of VR as stereotypes are *the girl in funny swimsuit with a headset*, but aside the headmounted displays and games hype, there are serious activities using other kinds of peripherals, some expensive like a CAVE (projection on walls of a room), some more suited to exploratory science like BOOM (Binocular Omnidirectional Orientable Monitor, a free 3D floating monitor with handles), and specialised assembly like simulators. Related activities are also Telerobotics/Telepresence, and Augmented Reality (superposition of images of real and virtual worlds, the challenge being to make them coincide).

Hard problems for VR include matching human sense astronomical bandwidth with limited peripherals functionalities and bandwidth: VR is mainly a graphics activity (limited to a small view field) and secondarily other sources are added, like sound (in progress), force feedback and other electrical sensitive interfaces. The graphics rendering is the (relatively) easy part, with new high speed graphics cards and good toolkits. Since VRML [18] is more and more used as a scene description language, in input mode for description of CAD-CAM models or as an output medium like in a 3DSciRendering system, and can be used in a distant/distributed way [15][17][18] the idea comes that behind the tool (the language), there may be more to share between the worlds of VR and ViSC.

2.2 Tools

VR tools are mainly integrated applications, developed specifically (e.g. for nuclear power-plant pre-visit planning, CAD review of petroleum refinery plant or cars design), or more appropriately, they are toolboxes allowing engineer to (relatively) quickly and easily use such applications. Current well known commercially available systems include Division dVise, Sense8's WorldToolkit, Superscape, and Medialab Clovis to name a few. Some freeware/shareware toolkits are also available, such as DIVE or MRtoolkit [8].

Beside the use (as input) of the scene description languages where VRML 2, a.k.a Moving Worlds is the one-up contender, those toolboxes are more and more object-oriented; most of them also include some *agent* or *reactive object* notion, verging on the boarder of AI. Those notions are used for example to model scene actor (or objects) scripts or constraints, without adding too much complexity in the modeling program. But a lot of the toolboxes are conceived with the *real* world in mind, not with some abstract or complex physical one such as we may encounter in the ViSC realm.

2.3   Philosophical and paradigmatic considerations

Data Mining is a kind of scientific-oriented activity that may considerably benefit from VR tools and concepts, and some emerging applications are aslready demonstrated [2]. But the vast majority of these examples are mainly for business-related activity, or at the least statistically oriented (such as principal components analysis/visualization), which is a small part or current ViSC activity. We can nevertheless extract some ideas from this experiment: the need for non nature-like VR worlds, and the multidimensional aspects. A secondary approach is the notion of interactivity with (scientific) object in the scene, such as property interrogation or hypernavigation to related information, which was not natural in the ViSC practice.

A lot of questions and discussions were around this notion of real-virtual world, and what were the required *qualities* to have a working VR environment for ViSC. If current ideas are that the principal quality for a ViSC render is the fidelity to data, there are some problems such as the conceptual universe that physicist are used to: how do we navigate in a fluid mechanics world (inside stream-lines?), and the use of new dimensions (eg, colorization or sonification of data) which is already in large use, is a deviation from strict realism. Taking into account such examples as geographic or city map reading, or the universe of air traffic controllers, or the experiments in relastivistic theory visualisation we see that what is in fact important is the accepted paradigm for the community of users.

What can we hope from such tools and new uses? Suggested possibilities range from discovery of new phenomena hidden in data (eg turbulence high level order, or shockwave in fluid data, ...), to refinements of models and visualization of phenomena dynamics, going even to experimenting new physics rules in a virtual simulated world. But the most immediate reward would probably, as we would have the concept of a virtual experience world, the sharing of this experience world as a way of collaborating and sharing insight (for both research and educational purposes).

Several working groups are active in this area today and maintain WWW sites on this topic [2][10][4][3][9]

3   Open issues and conclusions

Networking issues are important, especially in the WWW context. The kind of infrastructure used, network performances, and security constraints will for example be different if applications are used locally (Intranet) or on the Internet. Presently available bandwidth over Internet is limiting a lot usage of distributed visualization environments, but this does not apply to local area networks. All advantages provided by WWW technology can therefore be already used by many users.

Hardware availability concerns are more important in the VR context. On the immediate issues, availability of (relatively) cheap 3D hardware based on PC and 3D cards could change the current paradigm of VR, still perceived as a hard science necessitating huge investment. However the *VR station* will not be used if the peripherals are not efficient or difficult to use. Usage of VR will also largely depend on the added value for the users. This is were large enhancements are still needed. Not only graphical representation of data need to be significantly different in VR from what is used on *classical 3D* workstations, but GUI and ergonomics problems still need to be more deeply studied: navigating and handling object in a 3D world is not an easy task.

On a software point of view, both WWW and VR worlds still appear to be in a raising phase, with development tools, languages and APIs far from being stable and mature. Just considering the common 3D format in both world (VRML) gives a good view of this. It has proved from its first version a very high potential and convinced many future users or application developers, but much work still has to be performed on many aspects. Compactness is only one aspect and for example the set of available graphic primitives and attributes still limits its usability in all domains.

In the WWW world, Java language and associated object classes, even if they are already very powerful are still far from being rich enough, and efficiency remains an issue (probably not on a long term). The same applies to the different protocols in which many weakness still exist.

Collaboration between people with WWW and/or VR applications also appear as an important common direction. Browsers provide more and more facilities for allowing people to work together (groupware tools, audio/video conferencing), and some work already appear for helping in the development of collaborative interactive applications (see [12] for example). VRML developments also plan to incorporate collaboration features. All this is still in a preliminary stage and needs some time to mature.

However VR and WWW have already convincingly demonstrated their high potential and applicability in Visualization Systems and will provide in a near future powerful tools for scientists in their data analysis.

References

1. Crystalpacking visualization by vrml.
   http://dta.med.harvard.edu/ubc/banff/xpack.html.

2. Downloadable software from the geometry center.
   http://www.geom.umn.edu/software/download/.
3. Eurographics working groups.
   http://www.eg.org/WorkingGroups/.
4. Le groupe graphiques et interface homme machine d'aristote.
   http://www-aristote.cea.fr/gihm/ (in french).
5. Micado: Database for microbial genomes.
   http://138.102.88.140/cgi-bin/genmic/madbase.home.pl.
6. Npac visible human viewer.
   http://www.npac.syr.edu/projects/vishuman/VisibleHuman.html.
7. Opengroup desktop technologies: Xwindow system.
   http://www.opengroup.org/tech/desktop/x/.
8. Virtual reality page.
   http://www-ece.engr.ucf.edu/ mav/VR/vr.html.
9. Vrml data visualization tool.
   http://skynet.ul.ie/ keith/fyp/index.html.
10. The vrml repository.
    http://www.sdsc.edu/vrml.
11. W3c main page on graphics formats for the www.
    http://www.w3.org/Graphics/.
12. J. Beogle, C.A. Struble, and C.A. Shaffer. Leveraging java applets: Towards col-
    laboration transparency in java. *IEEE Internet Computing*, 1(2):57–64, 1997.
13. M. Grave. Distributed visualization in flow simulations. *Computers & Graphics*,
    17(1):9–14, 1993.
14. R.B. Haber and D.A. McNabb. Visualization idioms: A conceptual model for sci-
    entific visualisation systems. In G. Nielson et al., editor, *Visualization in Scientific
    Computing*. IEEE Computer Society Press, 1979.
15. M. Jern. Information drill-down using web tools. In W. Lefer and M. Grave, editors,
    Visualization in Scientific Computing, Springer Computer Science. SpringerWien-
    NewYork, 1997.
16. R. Orfali, D. Harkey, and J. Edwards. *The Essential Client/Server Survival Guide*.
    John Wiley & Sons, 2 edition, 1996.
17. J. Trapp and H-G. Pagendarm. A prototype for a www-based visualization service.
    In W. Lefer and M. Grave, editors, *Visualization in Scientific Computing*, Springer
    Computer Science. SpringerWienNewYork, 1997.
18. J. Walton and D. Knight. Presentation at worldmovers conference, san francisco.
    avalaible at http://www.nag.co.uk/visual/IE/iecbb/VRML2/wm1.html, jan 1996.

Information Drill-down using Web Tools

Mikael Jern

Vice President Technology
AVS/UNIRAS

Abstract The paper reviews the Information Visualization and interaction techniques needed to add another dimension to surfing the Web, *Information drilling* and *interactive data querying*, sometimes also referred to as *Visual Data Mining*. Information Visualization can be used to explore relationships by *drilling down* and retrieving more data within a region of interest in the visualized data, combining data mining, direct manipulation and data visualization with 3D Web tools. It is now possible to create desktop visualization applications that let users interact with databases with larger datasets over the network using both 2D and 3D interaction metaphors. The VRML standard allows users to view and navigate through 3D information data worlds and hyperlink to new worlds. Information drilling based on HTML's Image Map, VRML's anchor node and multiple predefined viewpoints will be explained and demonstrated. The image map in 2D and 3D graphics objects (glyphs etc) will represent the Visual User Interface to the information stored in the database. Also the advantages of using distributed component techniques based on plug-ins, Java Beans and ActiveX providing client-side data manipulation will be reviewed and illustrated. Over the next couple of years, we shall see 3D visualization evolve in giant steps into interactive data drilling on the Web providing visualization technology closely integrated with the data warehouse and multidimensional abstract and geospatial data models.

1 Information visualization

Information visualization can be used to explore and analyze relationships in multi-dimensional, large datasets by *drilling down* layer after layer and retrieving more information within a region of interest. The graphics in each layer can distinguish types of information by color, height, size, pattern, outline, texture, arrows or shapes, to name but a few. The data in figure 1, for example, is represented by outline (geographic), height and color.

It is now possible to create desktop visualization applications that let users interact with large datasets on the Web using fully 3D interaction metaphors - This is also the foundation for *Collaborative visualization - Share my data and design*. This paper considers the potential for information visualization to bridge the gap between the abstract, analytical world of the digital computer, and the human world. Concepts like *publishing, interactive 3D Web browsing, interactive data querying, data surfing, drill-down*, etc will be explained.

2 "Thin" versus "fat" visualization clients

The widespread popularity of Web technology has created a new information visualization technology model, in which browsers enable the widespread distribution of information using standard HTML techniques. The explosive growth of the Web has changed user expectations concerning the delivery of information to a client. The traditional information visualization approach allowed any client to communicate as a peer to any available server. The GUI model was either written in Motif for UNIX or Windows for the PC desktop. The Web introduces a new model in which the client GUI, based on HTML, is less functional and relies upon the data or application servers for visualization traditionally executed on the client.

In the Web-enabled world, the client is effectively reduced to a browser (viewer) of information supported by a server. A true Web client is not capable of program execution unless the executables are downloaded to the client as either Plug-ins or Components. This client is normally referred to as the *thin* client. A thin client, by definition, have minimal software requirements necessary to function as a user interface front-end for a Web enabled application.

Local data manipulation, information drill-down technique, context sensitive menus, object picking and other interactive user interface functions that traditionally have been available on the client are now controlled by the visualization server. In the *thin* client model, nearly all functionality is delivered from the server side of the visualization engine while the client perform very simple display and querying functions.

The most appealing aspect of the *thin* visualization client to information visualization users is that the overall cost of software and maintenance can be dramatically reduced. The *thin* client allows the application developers to eliminate the notion of software distribution at the client level (no license issue!), eliminate the notion of maintaining local software and supporting multiple operating systems on remote clients.

The concept of a *thin* client, however, raises the issue of client vs. server data visualization rendering. The standard Web browsers are *static* and do not permit any visual data manipulation at the client side. The user interaction is dependent upon the network bandwidth. Partitioning the visualization process between clients and servers is an effective way to distribute the computing resources. The most flexible visualization system allows the application developers to control the visualization partitioning.

Java is now being used to overcome some of the limitations. Java allows the creation of components *applets* or *JavaBeans*, which are automatically downloaded and executed on the local client. These components can significantly increase the data interaction between the client application and user, and allow tasks to be executed on the client. Java applets, for example, are interpreted on the client by the Java Virtual Machine, which is usually embedded in the Java-enabled browser such as Netscape's Navigator or Microsoft's Explorer.

These Java applets that deliver locally available executables are, however, still dependent on the network bandwidth. Depending on the scope and application, Java applets and its data sets must be downloaded. Java applets are only resident during execution and are therefore removed from the local disk after the completion of the task. As the demand for larger applets and data sets grows, significant download time could be incurred and the network becomes the bottleneck. Keeping commonly used applets resident on the client would significant reduce download time, although this practice is counter to the Java applet architecture.

The most compelling reason for the use of an *intelligent* client, such as a Plug-in or Component in a visualization application is the need for sophisticated user interface and data manipulation. Ease-of-use is often the primary factor for considering a Web-based solution. However, the current limitations with HTML and Java class libraries make the implementation of complex visual interface front ends very difficult. An intelligent client offers the opportunity to deliver highly graphical, highly interactive user interfaces that provide point-and-click navigation through multidimensional data models, such as exploring complex Data Mining trends and subsetting dimensions in an OLAP environment.

An *intelligent* visualization client provides local functionality through Plug-ins or Web components (ActiveX or JavaBean). Visual data manipulation is provided at the client side through locally stored components. Highly interactive user interface tasks are delivered that provide point-and-click navigation through multidimensional data structures. Visual data interfaces such as information drilling, moving a cutting plane through a volume data set etc are supported. Clearly, a full-featured visual data manipulation has many advantages over the rudimentary offerings of Java applets and HTML query forms.

Color plate 1 shows an example of an intelligent visualization client produced with AVS' GSHARP Web Edition. The 2D contour map, color legend, and the two charts to the left were produced at the server side with AVS' Gsharp using Java2D graphics imbedded in a Java applet. The special Gsharp Java Profile applet was transferred from the server and executed at the client side doing the profile calculation and drawing of the horizontal and vertical profiles.

3 Information drill-down on the Web

It is now possible to create desktop visualization applications that let users interact with databases with larger datasets over the network using both 2D and 3D interaction metaphors. The following reviews the Information Visualization and Web interaction techniques needed to add another dimension to surfing the Web; *Information drilling* and *interactive data querying*, sometimes also referred to as *Visual Data Mining*. Web-based Information Visualization can be used to explore relationships by *drilling down* and retrieving more data within a region of interest in the visualized data, combining data mining, direct manipulation and data visualization with either 2D or 3D Web tools.

3D interactive graphics on the network requires a 3D interactive format and a navigation system that combines the 3D input and high performance rendering capabilities. Virtual Reality Modeling Language, VRML, (pronounced *vermal*) is the language for describing multi-user interactive simulations - virtual worlds networked via the global Internet and hyper-linked within the Web.

VRML is an open, platform-independent, standard file format for 3D graphics that grew out of Silicon Graphics' object-oriented Open Inventor 3D toolkit in 1994. By defining a new file format to represent 3D scenes, and by creating stand-alone *client* viewing programs for that file format, today's Web browsers can also handle 3D scenes on the PC desktop platforms. VRML introduced in early 1995, is now driven by an *open*, growing consortium, which is quickly broadening its horizons and future development:

VRML 1.0   *Geometry*   1995 3D object representation

VRML 2.0   *Behavior*   1996 Put the static geometry into motion

VRML 3.0   *Sociality*   1998 Interface for multi-user interactivity

VRML's *anchor node* and *multiple predefined viewpoints* allow the users to view and navigate through 3D information data worlds and hyperlink to new worlds. Information drilling can also be implemented in 2D based on HTML's Image Map. The image map in 2D and 3D graphics objects (glyphs etc) will represent the Visual User Interface to the information stored in the database.

4 Image-map used to implement information drill-down in 2D

One of the most powerful uses of 2D graphics found on the Web is the Image map in HTML. Image maps are regions of your screen assigned to links. Clicking on one area of the image will take the user to one location, while clicking on another will take him somewhere else. The graphics that are displayed are just ordinary GIF or JPEG graphics. However, an additional file is kept with the image called a map definition file. This is an ordinary ASCII file that contains the definition of where the active *clickable* areas on the image are located. This information is stored as coordinate pixel locations with corresponding links. These areas can be defined as rectangles, circles and even arbitrary polygons.

In addition to the image and map definition file, an image map needs a Common Gateway Interface (CGI) script. This script is a special program that acs as a middleman between the browser and the map definition file. When the user click on the image map, the CGI script looks in the map definition file to see what to do. It then points the browser in the right direction and the user is transferred to another site.

A typical Image map definition file contains several pieces of information for the server's CGI image map routine. The contents include lines starting with words like rect, line and polygon, which represent the type of areas being defined.

```
<IMG SRC=«graph.gif» USEMAP=#map1>
<MAP NAME=«map1»>
  <AREA SHAPE=RECT COORDS= «0,0 100,100»
    HREF= «/cgi-bin/map.cgi?pick=upperLeft»>

  <AREA SHAPE=POLYGON COORDS= «
    262,231
    263,231
    ...
    203,169»
    HREF=«/cgi-bin/map.cgi?pick=DEN»>
</MAP>
```

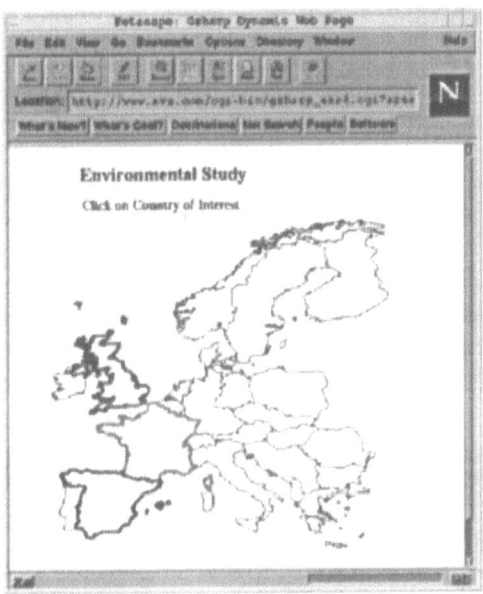

Figure 1: Image maps (countries) generated by Gsharp's Web Edition

GSHARP Web Edition from AVS supports Information Drilldown based on Image maps. GSHARP's object-oriented architecture provides the framework for automatically generated Image maps. The countries in Figure 1 were generated by Gsharp and are *clickable* areas with hyperlinks attached to them.

5 Dynamically created visualization

The VRML files can be either static or created dynamically:

- The user can view a *database* of already existing static VRML scenarios. The user interacts with a single VRML file at a time in a 3D browser. Any such

14

selection results from a server being contacted to deliver the appropriate VRML file.

- The user sends a request (with a HTML generated *form*) to the server and receives the selected static VRML scene from a database.

- Virtual VRML scenarios defined on the basis of a simulation or any analytical expression are generated interactively with CGI scripts. The HTML user interface form permits the user to control the visualization method and its attributes dynamically.

The CGI (Common Gateway Interface) is a powerful mechanism for transferring images and data that has been produced interactively over the Web. The architecture of an Application Visualization Server includes a number of parts:

- User Interface implemented with HTML forms

- Web browser Application Server which utilizes a Visualization Engine for batch graph generation

- CGI Handler Program controlling the dynamics of the Web application

- VRML browser for viewing the resulting visualization

The interaction between the Visualization Server software and the client *Web browser* is described in the diagram in figure 2. The visualization Web interface is created with standard HTML, which provides limited tools for designing a user interface called a *form*. This form contains a number of fields to be set by the user, which controls not only the data to be visualized, but also the visualization attributes. The user accesses the application server through this HTML page in a Web browser.

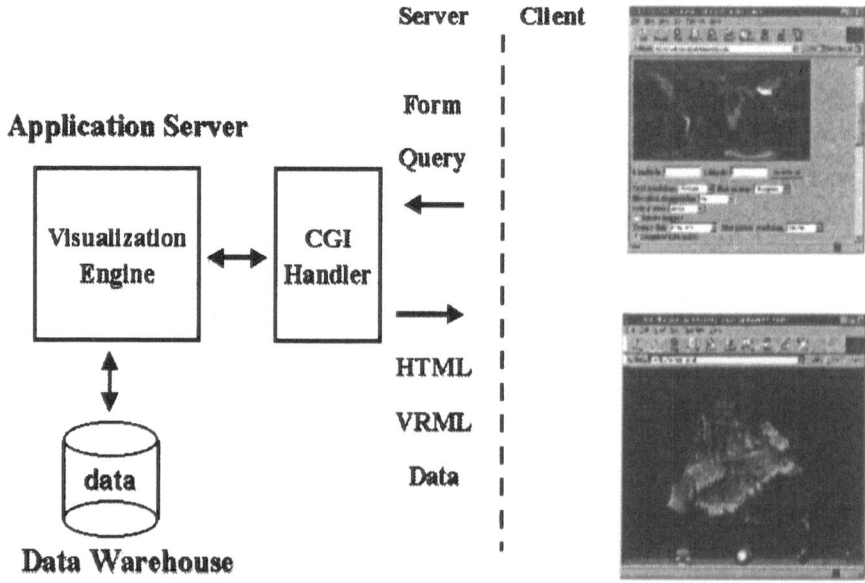

Figure 2 : A diagram of a *dynamic* Web VRML application

On submission of the form a Common Gateway Interface (CGI) script is executed on the Web server machine. The script contains the sophistication necessary to guide the visualization software in producing appropriately laid out graphics and the attribute information and data request specified in the form. This is then passed to the visualization server where it is used to set parameters in the Visualization Engine. The requested data is accessed, the map instructions are executed and the geometry is created, which is finally converted into the standard VRML 2.0 file format. The VRML is transferred back to the client and the appropriate VRML browser.

6 Multi cameras - Guided tours

The virtual camera motion is an important form of 3D interaction in information visualization. The navigation and visual experience on low-end machines use *guided-tour* and *point-and-click seek* navigation tools, techniques that are proven to be almost as appealing and effective as *free roaming* on powerful graphics workstations. The visual cues provided by interactive 3D viewing with continuous control offer invaluable help in understanding the represented data. If images are rendered smoothly and quickly enough, an illusion of real-time exploration of a virtual environment can be achieved as the simulated observer moves through the model. For example, in information visualization, large multi-dimensional data sets can be inspected and better understood by the user by walking through the 3D virtual data projections.

An example of these navigation tools is the *multiple virtual camera* feature in a VRML file. A number of cameras can be specified in the VRML file and are listed inside a Switch node. This allows the browser to give the user a selection of viewpoints. The user clicks on the *Viewpoints* menu item and a list of the available camera positions will pop up. The camera node can also be assigned a name, which can then be referenced as an *entry point* of the virtual world. Cameras can be placed in the object hierarchy just like any object or light and are therefore affected by the transformation nodes.

The multiple cameras feature is supported in a very professional way in most 3D browsers, color plate 2 shows a guided tour with the WebSpace 3D brower). By selecting a new viewpoint, the user is taken on an animated tour from the existing viewpoint through the landscape to the new selected viewpoint. Such a *guided tour* is useful, to guide the user through a complex scene and highlight special points of interest. The camera positions affected by the transformation nodes, allow the modeler to zoom-in at selected objects. These predefined camera positions are interactively specified by the modeler and stored in the VRML file.

7 Attaching data attributes to graphical objects

The effectiveness of interactive 3D viewers for communicating information about 3D environments can be dramatically enhanced by attaching annotations to the 3D scenes. Links in VRML works in precisely the same way as they do within HTML, thus pointing to an object with a link, will first highlight the object *visual cue* and if demanded bring up an application or data attribute that is designed for the selected object. These links can be used to develop information drilling in a 3D space. The WWWAnchor node in VRML provides the framework to have links to other worlds, animations, sound and documents.

Two ways of attaching attributes to a graphics object:

- The WWWInline node represents one of the most powerful features of VRML and is used to load additional VRML files from elsewhere on the Web into the scene. With this feature, large and complex scenes can be composed from a *repository* of smaller objects.

- The WWWAnchor node is the equivalent to the anchor tag in HTML. Hence, it represents the Web hyperlink. You can create an anchor to anything in the Web, to a Web text, a movie, or another VRML world.

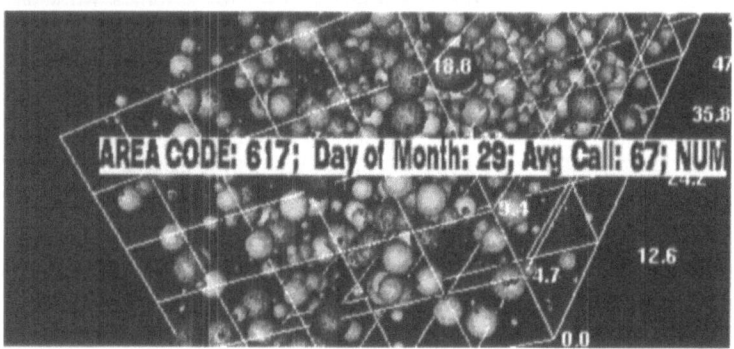

Figure 3: Interactive 3D Information Visualization on the Web using the WWWAnchor Node :
«http://www.tel.com/call.htm»
  description
«AREA CODE: 617; Day of Month: 29; Avg Call; 67; NUM .....»

8 Five dimension display - "Glyph visualization on the Web"

A 3D scatterplot display uses *glyphs* to represent multi-variate data, where each characteristic is determined by the data. Each data point is represented by a sphere or *bubble*. The position (x,y,and z), size, shape, and color of the sphere can each be used to represent a variable in the data, an example of a multi-variate display in five dimension. The user can also view attached additional abstract information of a selected sphere.

The glyph visualization helps the user to see trends in very large data sets. However, the user is not limited to an overview of the data, but can retrieve full and rich descriptions of underlying database attributes. Only showing details when they are requested is vital for the concept of dynamic queries. The user applies *details-on-demand* by pointing at a glyph object and clicking the left mouse button. The object is high-lighted to indicate that it was selected. The client creates a query that is sent to a server.

The commercial application in color plate 3, produced by a prototype application developed with AVS/Express software, uses the *glyph* display technique. The example shows the possibility to construct a dynamic query system *information drill-down* to a database using HTML forms, VRML and a CGI script. The application provides controls to choose a data set, select a variable of interest, and navigate among the dimensions of the variable.

An SQL query, given in an HTML form, is sent to the Web server and the user receives a 3D glyph display in VRML format in return. The user then interacts in 3D space by pointing directly to a *glyph* (sphere in this case). Each glyph is represented as a WWWAnchor Node, which hyperlinks the object to additional information about the selected data. This *drill-down* technique provides an immediate graphical representation of data and its attributes, without the intervention of agents like 2D controls (scrollbars, and other traditional direct manipulation tools), or 3D widgets that are separate from the data representation. By allowing users to interactively recall and view the attached information by selecting objects of interest during navigation, the interactive 3D viewer becomes a natural front-end for dynamic querying of information.

9 Viewing vs. client application plug-ins

Despite new types of information, developers and information sources still cling to the standard HTML, GIF and VRML formats for a simple reason - it's practically guaranteed that the widest Internet audience possible can view the information. Real-time visual data manipulation, however, doesn't translate well into these standards. While the VRML file format allows distribution of visualization scenes to the Web, the user has no interactive control of the actual underlying data. The *mapping* of numerical data into geometry format (VRML) takes place on the server side. The client *3D browser* can only navigate in the 3D world. Manipulations of the data and dynamic control of the visualization attributes that have been used in traditional data visualization systems take place on the Client side.

For example, the user wants to interactively slice through a 3D volume of data. The Visualization Engine at the Server side must generate a new VRML file for every new data slice. The user on the Client side clicks on the browser's Reload button every time to get a new updated VRML. Unfortunately, with low network bandwidth and maybe several hundred people clicking their Reload button every minute, the Web server would become overloaded and preventing anyone from getting images or data. Clearly, in some situations the wide acceptance of HTML can't offset the inherent limitations of the format. In these cases, another option is needed.

If you need to interact directly with your data in real-time, your information just can't survive a translation into HTML and VRML. The solution is to move part of the actual data rendering process from the Server to be imbedded in your Web browser at the Client side. Any of the following techniques are available: *Visualization Plug-ins*, *Helper Applications*, Java applets or ActiveX controls.

The most compelling reason for the use of local tasks is the need for sophisticated user interfaces. Ease of use is the primary factor for considering a Web based solution. However, the limitation of HTML makes the implementation of complex front ends very difficult. A more sophisticated Client can deliver highly graphical, highly interactive user interfaces that provide point-and-click navigation through complex data models and data drilling.

Java applets are still dependent on network bandwidth to deliver the executables. Depending on the scope and capability of the applets, large numbers of executables may need to be downloaded in order to accomplish the task. Executables are only resident during execution and are removed from the local disk after the completion of the task. As the demand for larger applets grows, significant download time could be incurred.

Plug-in modules are programs specifically written to run *embedded* within a particular Web browser. Netscape plug-ins are the most popular format and are now emerging as the standard. Visualization Plug-ins can be used to let the user (client) read a script language, which controls the visualization type, attributes and the data to be visualized. The Web browser knows about the Visualization Plug-in , and will automatically launch it and load the Plug-in into it once the data transfer finishes. The Visualization Plug-in will perform the data manipulation and rendering locally at the client side.

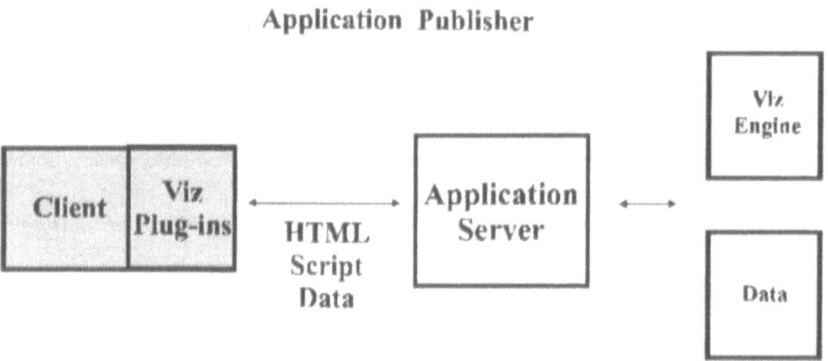

Figure 4: The Application Plug-in scenario. The mapping of data into geometry and rendering is performed at the client side. The user can interactively manipulate the data. A script language can be transferred together with the data to set up the appropriate visualization method. This special visualization Plug-in performs *data slicing* through a volume dataset..

There are, however, always trade-offs. the *application publisher* scenario is not suitable for transferring very large datasets over the Internet. Here the VRML scenario would be a more appropriate Web visualization method.

10 Client-side behavior with visualization components

The ultimate Web-based visualization capabilities will be delivered through Web components. Visualization on the Web will become even more active and dynamic, when JavaSoft's JavaBeans and Microsoft's ActiveX visualization components begin streaming down the Internet to the Web browsers.

In many applications, sophisticated components are increasingly becoming the main feature, providing capabilities now that would take a considerable amount of development time and expertise to code. Developing some of the intricate components that are needed in interactive Web applications on the market today can be prohibitively expensive and require deep knowledge.

ActiveX essentially extends Microsoft's existing and proven Object Linking and Embedding (OLE) and Component Object Model (COM) technologies to the Web and is optimized for the Windows environment. Microsoft's Internet Explorer (IE) 3.0 is the first major application to exploit the ActiveX technology. ActiveX components can be used as *plug-ins* or you can create applet-like programs (programmed in Java!) embedded inside HTML pages that use <OBJECT> tag to refer to components *ActiveX controls*.

AVS/Express is a visual component development framework, in which developers design, build and customize visualization components using a Visual Programming technique. These components are either C++ classes, ActiveX or Plug-ins that can be combined to create Web applications. Sophisticated components that do everything from displaying information in sophisticated 3D and 4D scientific visualization techniques, to providing real-time graphs and images on everything from medical MRI images to sensitive financial information.

11 Conclusion and future trends

In this exploding world of abstract data, there is great potential for information visualization to increase the bandwidth between us, and an ever growing and ever changing world of data.

The future trends and improvements in Information Visualization for the Web can be summarized:

- *Information Drilling* on the Web
- 3D visualization on the PC desktop
- Non-immersive VR navigation using *Visual* User Interface technology
- Real-time visualization of very large data sets
- Components - JavaBeans and ActiveX
- Collaborative *multi-user* visual environment - Guided analysis
- Web database visualization - Visual Data Mining

Over the next couple of years, we will see VRML evolve in giant steps into interactivity and multi-user participation based on the new emerging standard VRML 2.0 and the future evolving more collaborative VRML 3.0. Visualization will develop into interactive data drilling on the Web providing visualization technology closely integrated with the database.

As the technology of information visualization becomes more important to the decision-making process, it will have a natural tendency to migrate to the Web so that it can be made available to the largest number of users. The VRML virtual camera motion feature provides users with an animated tour of the data. Glyphs can be used to construct dynamic query systems that operate on the Web. Finally, plug-ins, Java applets and ActiveX controls move the data rendering process to the client side to overcome bandwidth limitations. Each of these technologies is contributing to the rapid transfer of information visualization on the Web.

Color plate 4 shows an example of an Application Plug-in prototyped by AVS/Express to perform Volume Rendering at the client side. Compared to general-purpose VRML and http, the Application Plug-ins allow more sophisticated interaction between the client application (the web browser plus plug-in) and the visualization server. It supports direct manipulation of both data and the visualization parameters. For example, dynamic specification of data/color mapping is accomplished using specialized interaction tools.

References

1. Brown, J., Earnshaw, R.A., Jern, M.,Vince, J.A., Visualization, Using Computer Graphics to Explore Data and Present Information. John Wiley & Sons, New York, 1995, ISBN 0-471-12991-7

2. Earnshaw, R.A. and Vince, J.A., Computer Graphics, Developments in Virtual Environments, Academic Press,1995, ISBN 0-12-227741-4

3. Gross, M., Visual Computing The Integration of Computer Graphics, Visual Perception and Imaging, Springer, 1994 ,ISBN 3-540-57222-8

4. Woo, D., Cole, The World Wide Web Book, 1995, Springer, ISBN 0-387-94433-8

5. Abraham, R., Jas, F., Russell, W., The Web Empowerment Book, Springer 1995, ISBN 0-387-94431-1

6. Pfaffenberger, B., Publish it on the WEB, Academic Press, 1995, ISBN 0-12-553140-0

7. Scateni, R., van Wijk, J., Zanarini, P. (editors), Visualization in Scientific Computing '95, Springer Wien, 1995, ISBN 3-211-82729-3,

8. Veltkamp, R.C., Blake, E.H. (editors), Programming Paradigms in Graphics '95,Springer Wien, 1995, ISBN 3-211-82788-9

9. Gobel, M. (editor), Virtual Environments '95,M., Springer Wien, (1995), ISBN 3-211-82737-4

10. Gobel, M.,   Mlller, H.,   Urban, B. (editors), Visualization in Scientific Computing,Springer Wien, 1995, ISBN 3-211-82633-5

11. Herzner, W., Kappe, F. (editors), Multimedia/Hypermedia in Open Distributed Environments, 330 Pages, Springer Wien, 1994, ISBN 3-211-82587-8

12. Grave, M., Le Lous, Y., Hewitt, W.T. (editors), Visualization in Scientific Computing, 218 Pages, Springer Wien, (1994)

13. Tufte, E. R. 1990. Envisioning information. Cheshire, CT: Graphics Press.

14. December, J. and N. Randall. 1994.

15. Ford, A., Spinning the Web, How to provide Information on the Internet, Thomson Computer Press, 1995, ISBN 1-850-32141-8

A Prototype for a WWW-based Visualization Service

J.C. Trapp, H.-G. Pagendarm

Deutsche Forschungsanstalt für Luft- und Raumfahrt
Bunsenstr.10, D37073 Göttingen, Germany

Abstract. A WWW server accepts data from a user in order to produce a visualization of the data. The user can control the visualization algorithm over the network using a JAVA-based user-interface. The visualization is transferred back to the client as a VRML world file and is pushed to the VRML Viewer. The prototype solves principal technical problems and serves as a feasibility study for an open service provided to WWW users and a platform independant access to visualization methods.

1. Rationale for a WWW-based visualization service

In recent years visualization has rapidly evolved. A variety of visualization techniques has been developed and implemented. A number of general purpose visualization systems provide good support for users of visualization as a tool for their daily work. Such systems usually comprise a suite of the most popular visualization methods. As new and advanced visualization techniques are being developed they are obviously not implemented in every visualization package. Indeed the scope of a given system is usually quite limited. Some systems try to overcome this dilemma by using a modular architecture which allows the user to extend the system at his own needs. For example, a large number of additional modules have been created for the AVS-system over the years by AVS users and many of these modules are even available in the public domain for others to use [6]. The AVS approach is quite successful, but the availability of visualization methods to the general public still remains limited since an AVS licence is required and certain hardware restrictions apply.

1. 1. Demand for general availability of visualization techniques

There is a demand to make visualization techniques available for general use. This demand is driven by various clients. A low-cost or even no-cost visualization service on the WWW would drastically lower the entry barrier for the use of visualization. In particular, for users who rarely need visualization the costs or overhead created by network usage may be far less than what is needed to buy and maintain an in-house system.

Frequent users of a given visualization system may still have a strong desire for additional visualization service. Since the scope of methods available in a given system is limited, additional methods may be used remotely and the results may be combined with those results obtained locally. This may apply for highly specialized methods which typically cannot be found in commercial software packages.

Error estimation is another major reason to look for additional visualization methods. In today's software implementation, the various sources of errors are not very well treated. Detailed information about the implemented algorithms is often not available. Therefore, it is good practice to visualize a particular feature of interest in a set of data by using a variety of different and independent algorithms in order to verify the result and obtain an error estimate. In some case this can be done using a single general purpose visualization tool. There are cases, however, where the use of more than one system provides a significant advantage. Access to alternative systems over the WWW would come in handy. A good example would be to compare various algorithms for particle tracing.

The visualization research has good reasons to make new algorithms available to the public. Usually, visualization research does not directly lead to a commercial product designed to make profits. In most cases, advanced methods are first created and used in the scientific community. These developments are soon available to the public through published literature. Many obstacles prevent these algorithms from being rapidly spread and incorporated into a large number of systems as system development is in the hands of few organizations with scarce resources. On the other hand, application of algorithms is a key issue to future developments and improvement. Wide spread usage of visualization algorithms triggers cooperation and new ideas in the visualization community. Today this cooperation is slowed down by hardware and software constraints in the various research organizations.

Comparative visualization has attracted significant attention recently [10]. This often requires the combination of data from different sources and different visualization algorithms for a single image. While one part of the data is usually familiar one along with the methods that treat this part of the data, the other part of the data may only be used once for the sole purpose of comparison. For example, a researcher numerically simulating flows tries to match his result with some flow experiment done elsewhere. In such a case the appropriate methods to visualize the experimental data may not be readily available to him. Experience shows that direct cooperation between persons in different organizations will probably be required to reach the goal. However, such cooperation is subject to time constraints. Once researchers, who supply data also supply visualization methods on the network, these obstacles will vanish. The likelihood for actual use of research results would increase.

1. 2. Use of remote visualization methods

The world-wide-web WWW offers an excellent opportunity to make visualization techniques available to the general public. Designers of visualization tools can offer the use of their tool as a service on the internet. The necessary components to access the internet and the WWW are already available to the public and wide-spread. One may assume that skills to use such service have already been developed by the envisioned client. Internet search mechanisms are in place to draw the client's attention to such service as soon as it becomes available.

Basic components are available to allow the creation of a WWW visualization service. The JAVA programing language offers the required functionality to design the user interface and communication between the visualization client and the visualization server. JAVA is supported by the major WWW browsers and as such widely available. VRML offer the necessary support to deliver the visualization to the client for further use. As shown in this paper, the technical problems were solved and a visualization server was implemented. For the present time, this prototype of a visualization server is a no-cost service on the WWW since DLR is a non-profit organization. Much research is presently going into the development of save accounting methods on the WWW. Because of the high interest others have in the issue of pricing internet service, one may assume that solutions will soon be available that will allow for commercial visualization servers (see: [11]). This paper does not address this issue. Also, a detailed analysis of a potential security problem is beyond the scope of this paper.

Ideally, there would be a standard for data formats to be accessed by visualization algorithms. Such standard does not exist and a large variety of formats is in use. Experience shows that the data format as long as it is well described does not pose a major obstacle for cooperation. Most formats are easily converted to other related formats and the typical client of visualization service will have some programming experience and thus will be able to perform data conversion.

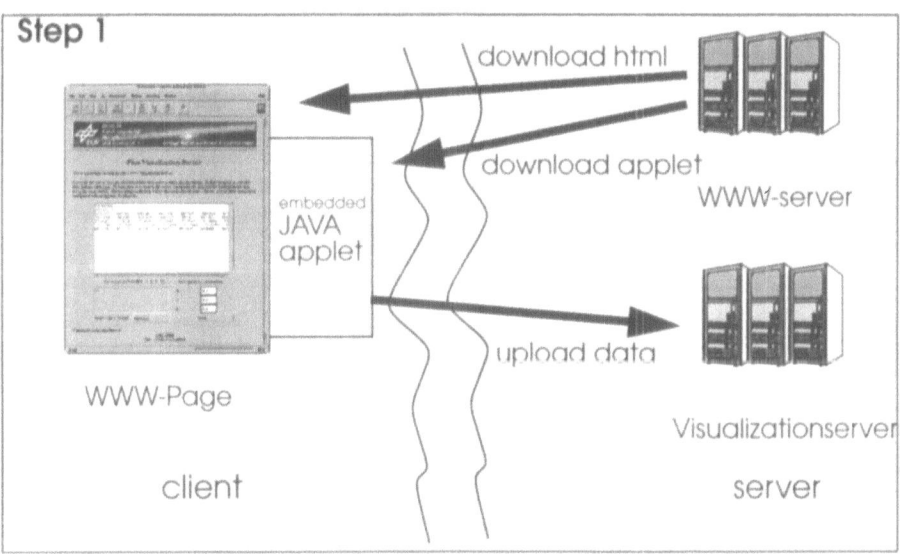

Fig. 1: Communication links between the client's site and the server's site (Initial steps). The html-page and the JAVA applet are downloaded first from the standard WWW-server to built the front-end at the client's site. Then the applet connects to the visualization server.

2. Architecture of the Visualization WWW server

2. 1. Accessing the server

The visualization server may be accessed as any other WWW-page using a browser such as Netscape. The home page of the visualization server contains the necessary information on how to use the server and offers description of accepted data formats. The home page, as well as the front page of the visualization server are supplied by the usual WWW server (Figure 1). In our case, we make use of the Apache daemon. When the user loads the front end of the visualization server a JAVA applet is loaded to the client site which drives the user interface and the communication between visualization server and client.

2. 2. User interface

The JAVA applet creates an interactive user interface which runs on the client's site (Figure 3). This user interface is presently restricted to functions required for the prototype application. However, the user interface may be as complex as necessary. The JAVA language offer a rich suite of methods for user interaction which were explored in earlier applications such as the interactive design of aerodynamic profiles[12], the Vectorize Tool [13] and the 3D geometry viewer [14]

2. 3. Loading user data

There are two ways to easily access user data without imposing extra programming effort on the user. Both methods were explored in earlier application. Both methods allow data access without effecting the usual security barriers of computer systems on the net.

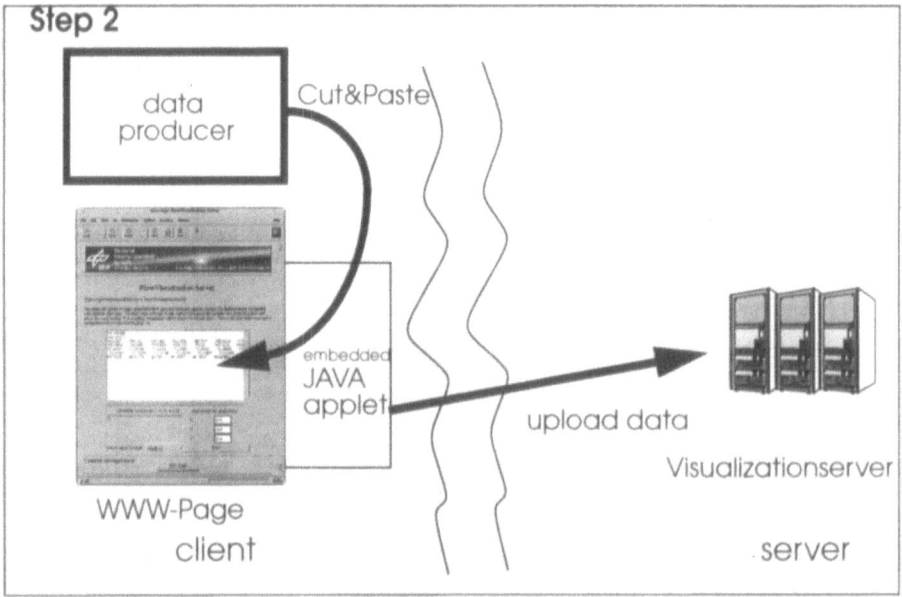

Fig. 2: Using the Cut&Paste or Clipboard function to enter data into the front-end

For a limited size of user data we implemented the Cut/Paste method. The JAVA driven user interface supplies a text window to input user data. It is obvious that the user would not use such a window to type in data by hand. In general, the user will own an application which will be data producer. This may either be a simulation program, a data base program or a simple texteditor. These programs frequently produce output to be seen on the screen. Provided this output has one of the acceptable formats, this stream of text may simply be handed to the JAVA front-end of the visualization server using the Cut/Paste function or the Clipboard function of the window system (Figure 2). This is an extremely simple way of entering the data. This method is well known to any WWW user. There are no considerations of access protection which would unnecessarily complicate the use of the visualization server.

A second method of loading user data is by using the WWW-URL address scheme. This method was also explored in the Vectorize Tool [13]. The method requires a little more effort on the user's side. In order for the method to work, the user needs to be able to make his data accessible on a Web-site.

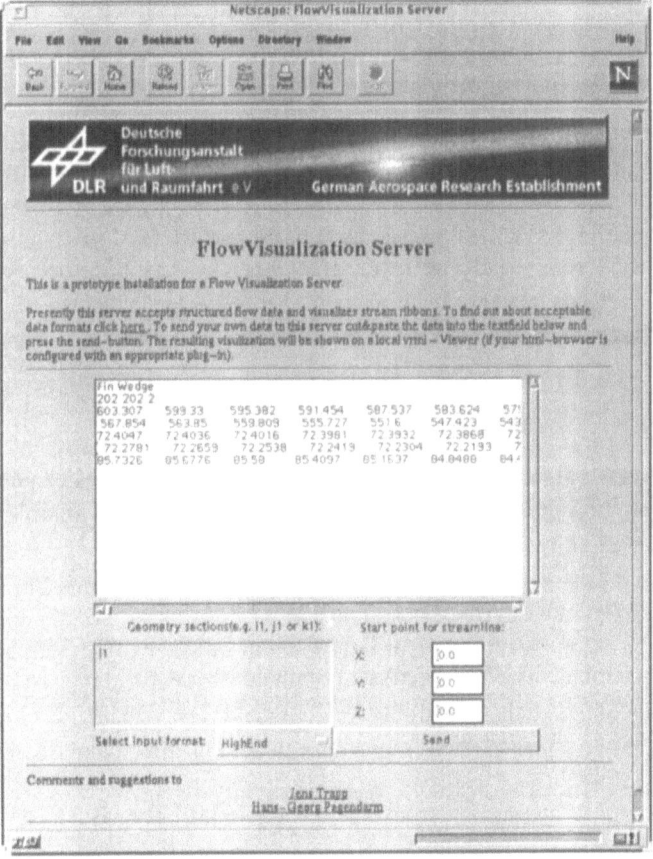

Fig. 3: User interface of the DLR WWW Flow Visualization Server. The user data is transmitted to the server by Cut&Paste into the white text field. Additional control elements such as menus and parameter input allow to control the visualization method.

The data needs to be supplied in a text-file in one of the specified formats on any Web site. The data may be read protected to the general public but must be accessible for the host running the visualization server.

The visualization server's user interface will query the location of the data as a standard URL address. This address will be use to load the data to the visualization server.

2. 4. Communication between visualization client and visualization server

Besides driving the user interface on the client's site, the JAVA applet has a second task. As soon as the user's input is collected and the user's data is accessed, the applet builds up a socket communication link to the visualization server. The visualization server is independent of the WWW server which was used to download the front-end to the client's site. A separate communication channel is used to send the data to the visualization server (Figure 1). After having received the data, the visualization server will create the visualization according to the controls set by the user in the front-end. After the visualization is created, the JAVA applet on the client's site is notified in order to initiate the downloading of the visualization.

2. 5. The Visualization server daemon

The daemon running on the visualization server site is implemented in JAVA as well. This JAVA server applet handles the communication with the client applet. It also passes the data to the appropriate visualization routines. Presently, only some demonstrators are implemented. However, any visualization routine which does not require permanent user interaction may easily be integrated into the visualization server. Such visualization methods need not be implemented in JAVA since binding to code written in other languages such as C or C++ is possible. It is also possible to pipe the data to separated processes which perform the visualization.

2. 6. Visualization methods

Many visualization algorithms are mapping techniques. They extract certain features or properties and map those to a visualizible object (the term visualizible object was introduced earlier by [4]). Example for such mapping methods are:

- selection of an index slice from a 3d volume on a structured grid. The visualizible object is a (coloured) grid of polygons.
- generation of an isosurface, and interpolation of scalars onto that surface. The visualizible object is a (coloured) set of polygons
- generation of a particle trace from a velocity field. The visualizible object is a line.
- generation of a shock surface from a density field [7], [8]. The visualizable object is a set of polygons
- generation of a time surface from a velocity field [10] The visualizable object is a set of polygons
- generation of a texture from a velocity field [5]. The visualizable object is a textured surface.

This is only a small number of examples. Modular systems like our HighEnd flow visualization system [9] or the GRTools [15] are easily plugged into the JAVA visualization daemon, as their mapping methods are typically implemented within a separate module. After collecting experience with the prototype visualization server we intend to make the suite of mapping methods of the HighEnd visualization system available through a visualization system.

By integrating the HighEnd functionality into a **WWW** visualization server we can overcome a major disadvantage of the present system. Today the HighEnd flow visualization system is equipped with it's own rendering module. In order to achieve high rendering speeds this renderer is dedicated to specialized hardware. In our case we built upon Sun Microsystems XGL graphics library to drive our Sun workstation graphics. Through the visualization server concept we can make the functionality of our visualization tools available to user of non-Sun hardware. With respect of present network speed we expect the clients for the visualization server primarily to be within the DLR network. This allows us to concentrate our efforts on solving the technical problems first and deal with commercial issues and security consideration and load balancing at a later time.

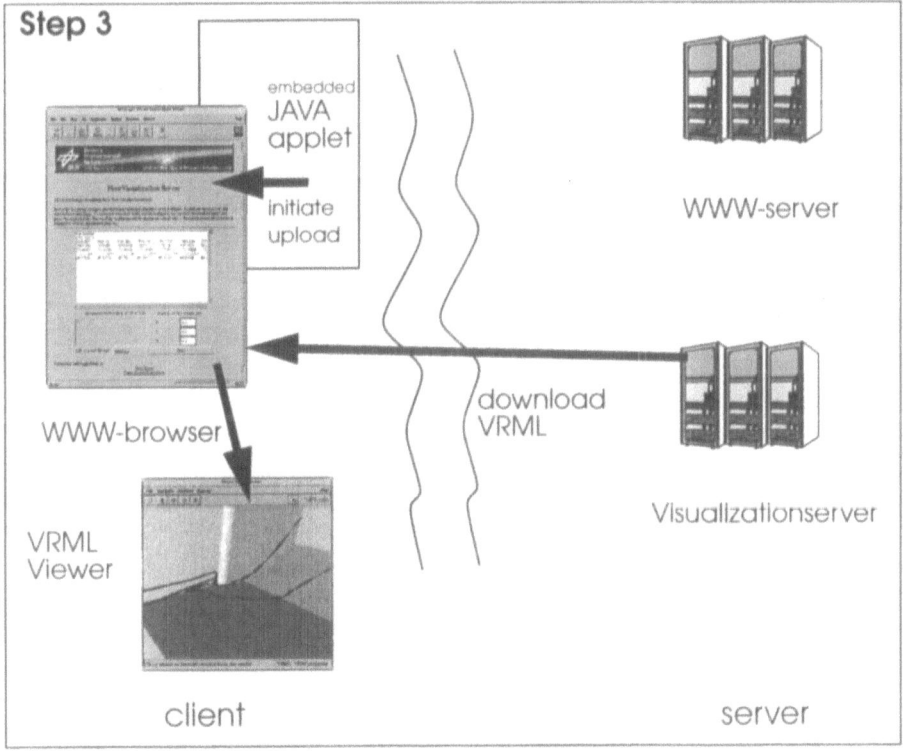

Fig. 4: Downloading the result. The front-end applet forces the WWW browser to download the VRML from the visualization server which is automatically passed to the VRML Viewer

2. 7. Creating the visualization

The visual objects as they are created by the visualization modules are converted to VRML worlds [1]. This approach has been suggested earlier by Wood et al. [17] who used VRML to distribute visualized environmental data. Wood et al. ran an IRIS Explorer based visualization system which was controlled via the WWW. The system is dedicated to allow access to environmental data which is provided with the system. Compared to the approach suggested here, one would say that the data resides at the server's site. The visualization is then exported as a VRML world. The system described here, however, allows the visualization of data that is supplied with the server as well as data from the user. Therefore, it has a much larger potential for a wide range of applications. It also allows the user to combine visualization from different sources in a single image.

We take an approach similar to Wood et al. only for transmitting the visualization. As soon as the VRML structures are created from the visualization pipeline the JAVA visualization daemon notifies the JAVA front end applet at the client's site.

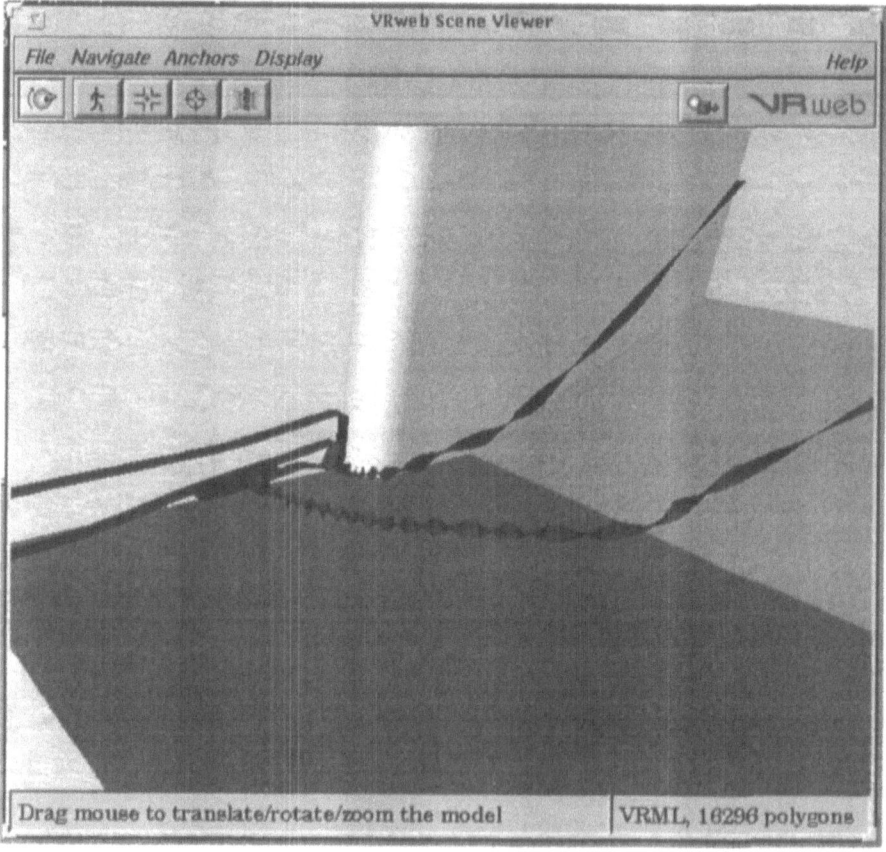

Fig. 5: Flow Visualization of vortices near a blunt fin transmitted to the VRML Viewer. The visualization may be interactively viewed using standard interaction functions of the VRML Viewer.

2. 8. Downloading the VRML world

Once the JAVA front end applet knows that the visualization has been created it will initiate the downloading of the visualization. Since the JAVA applet is running within a WWW browser it can force the browser to download the VRML world. This has the advantage that the visualization will be displayed in the environment defined by the user. Typically the user has predefined plug-ins attached to his WWW browser to deal with VRML world. Thus, the viewing and storage of the visualization downloaded from a WWW visualization server does not require any extra skills or installation on the user's side. The JAVA front end forces the browser to download a dummy URL with extension ".wrl" from the visualization server. Note that this data is inquired from the port of the visualization server and not from the WWW server since the VRML world was never stored as a file (Figure 4). This also ensures that only the sender of the data can retrieve the visualization of his data. The URL scheme is only employed to hide this special communication link from the browser, so that the browser will proceed in his usual manner to display the data once it is downloaded.

The URL get-procedure connects to a port of the visualization server daemon. The daemon then transmits the VRML world. It is important to note that the data is not stored outside the visualization server daemon. It is not written to disc or any other media. The visualization server acts as a simple pipe which receives user data and converts it into visualizable objects in a ".wrl" format. Once the VRML code is transmitted there is no copy of the user's data left at the visualization server.

2. 9. Displaying the visualization

Since the browser called for a URL with extension".wrl", it will automatically bring up the appropriate VRML browser according to its plug-in table (Figure 5). The user will then be able to view the visualization interactively on his own computer and save it to a file if desired.

3. Building complex scenes

In many cases, it will be desired to employ a variety of visualization methods on a set of data and build a complex scene [10]. This goal may be easily achieved with visualization servers. Since each visualization creates visualizible objects which is transmitted to the client as a ".wrl" file, these objects may be collected and combined at will by the user. VRML supports the combined viewing of several ".wrl" files with it's *WWWInline* command. In the simplest case, this may be done by manually creating a trivial file. Such a file consists of just a list of *WWWInline* commands which force the VRML Viewer to display all selected VRML worlds simultaneously. More complex tasks will be better performed with the help of VRML Editors which are readily available (e.g. [16], [2]).

4. Future developments

We feel that the prototype of a visualization server presented here has succesfully demonstrated the feasibility of a WWW based visualization service. Subsequently the technology was used to establish a permanent service at the DLR in Germany named the "Vis-à-Web-system". This system was used at the TU Delft as well to make available one of their visualization methods called spot noise textures. Further installations are exspected. After a period of testing we will make the JAVA client applet and the JAVA server applet available to encourage other people to use this mechanism to make their visualization methods available. We also intend to embed the major part of the modules today offered in the HighEnd visualization system into visualization servers and thus reduce hardware dependencies of our own visualization products. The concept turns out to be useful for a variety of other services beyond visualization as well. DLR will maintain a web page collecting all avilable installations of this service software system. The web page entitled the "Vis-à-Web-repository" may be found at http://www.ts.go.dlr.de/Vis-a-Web/ . This page will also offer the starting point for downloading the software.

5. References

Efficient and Reliable Integration Methods for Particle Tracing in Unsteady Flows on Discrete Meshes

Christian Teitzel, Roberto Grosso and Thomas Ertl

Computer Graphics Group, University of Erlangen
Am Weichselgarten 9, 91058 Erlangen, Germany

Abstract. In real applications the velocity field of a flow is not available in analytical but in discrete form. One goal of this paper is to analyze particle integration methods for discretized data defined on meshes with regard to numerical efficiency and accuracy. Careful error analysis of the particle tracing process relates the error of velocity interpolation in space and time to the error of the numerical integration. Hence, a fast integration routine which provides accuracy similar to that of interpolation is necessary. This leads to a robust integration routine with adaptive step size control and error monitoring. A second aspect of this work is the treatment of stiff problems. Stiffness occurs in flows with strong shear deformations or vorticity. To detect stiffness in a given flow field, the Jacobian of the velocity field is analyzed. Implicit integration methods are used to handle stiff systems of ordinary differential equations.

1 Introduction

Flow visualization tools based upon particle methods continue to be an important topic of research. With respect to numerical integration fourth order Runge-Kutta schemes are widely used within the visualization community, some times without a detailed error analysis. Especially, the interrelation to the ubiquitous linear interpolation of grid values is rarely discussed.

When computing a particle trace, an initial value problem for an ordinary differential equation has to be solved (see Section 2). If the velocity field of the flow is given in an analytical form, integration algorithms of high order are preferable like extrapolation methods or high order Runge-Kutta schemes. However, in real applications vector fields arise that are defined on discrete grids, since these velocity fields are given by numerical simulation or by measurement. For such rough vector fields higher order algorithms like extrapolation methods are useless.

In most particle tracing modules integration methods of the Runge-Kutta type of at least order four are used, e.g. the NAG-Advect modules of the IRIS Explorer use an adaptive RK4(5) scheme. Also Stalling and Hege use an adaptive RK4(3) algorithm in their Fast-LIC module [10].

In Section 2 we establish a relation between the errors introduced by the linear interpolation of the velocity on time and space domain grids and the errors introduced by the numerical integration. This leads to a robust integration routine with adaptive step size control of type RK3(2), which is not more accurate than necessary but significantly more efficient than the adaptive Runge-Kutta methods of higher order and the classical RK4 with fixed step size or the step doubling approach [1] or some other heuristic step sized adaption techniques [7].

In Section 3 the problem of stiffness is discussed. Mechanisms of detecting stiffness in a given flow field are analyzed and it is discussed in which data sets stiffness can be expected. Finally, implicit integration methods are suggested to cope with stiff systems of ordinary differential equations.

2 Integration Methods for Discrete Data

This section deals with the fast and accurate computation of particle traces in steady and unsteady flows. Lagrange visualization techniques of vector fields are based upon the numerical solution of an initial value problem for the following ordinary differential equation:

$$\frac{dx}{dt} = v(x, t) \quad , \quad x(t_0) = x_0 \tag{1}$$

where v denotes the velocity vector field, x the position , t the time variable and x_0 the start value at the initial time t_0.

Usually, the numerical solution for a particle trace is given on discrete grids $x_\Delta(t_i)$ at discrete time steps $t_i \in \Delta = \{t_0, t_1, \ldots, t_n\}$. The accuracy of a particle tracing algorithm is limited by the discretization error of the velocity field and the error of the numerical integration. If an integration method of order p is used, then the discretization error is characterized by

$$\|x(t_i) - x_\Delta(t_i)\| \le C\tau^p \tag{2}$$

where C is a positive constant, τ the maximal step size in the grid Δ and $t_i \in \Delta$.

Based upon this information, it is not possible to determine a good time discretization a priori and therefore equidistant time steps are usually chosen with $t_i = t_{i-1} + \tau$. However, it cannot be expected that grids equidistant in time direction solve problems with completely different characteristics appropriately. Even within a single flow, a different behavior of the flow field in different regions of the computing domain can be expected.

A reliable and efficient algorithm should be able to construct a problem adapted grid Δ. In this way the two goals, better performance and higher accuracy, can be achieved simultaneously. Hence, we are interested in an integration method which provides the desired accuracy with the minimal possible cost, including the generation of the approximating grid. This leads to adaptive integration algorithms.

In order to select an integration scheme, single-step or multi-stage methods, and multi-step methods are considered. In a single-step method like Runge-Kutta, the position $x(t_n)$ depends only on $x(t_{n-1})$. In a multi-step method, the results of a fixed number of previous time steps are used to compute the new position $x(t_n)$. The order of the method depends on how many previous time steps are used to calculate a new x. Multi-stage schemes are single-step methods with multiple function evaluations per time step, e.g. Runge-Kutta methods of at least order two.

Here we have to take into account that the overhead of a multi-step method for the computation of a problem adapted grid $x_\Delta(t_i)$ results in a higher cost than in the case of a single-step method. Because of this fact only multi-stage methods are considered in this text.

For adaptive integration algorithms the right-hand side of equation (1) has to be evaluated not only at arbitrary positions in space but also in time. In order to calculate the velocity v, tri-linear interpolation in space and linear interpolation in time is used. Higher order interpolation schemes in time would require to keep many complete time steps in memory. This would be a problem because of the huge data sets which usually arise from numerical simulations.

The interpolation scheme causes an approximation error of the velocity field. This error limits the maximal possible accuracy that can be reached by numerical integration of the particle trajectories. The error originating from the discretization of the velocity and the error resulting from the numerical integration are separately considered in order to estimate them.

If the discretization constants are h_t in time, the size of the time step given by the numerical simulation, and h_x in space, the size of the space discretization, then the error in the approximation of $v(x,t)$ resulting from the linear interpolation is of order $O(h_t^2)$ in time and $O(h_x^2)$ in space. In order to understand the influence of these errors on the integration, the exact trajectory generated by the velocity field $v(x,t)$ is compared with that generated by the vector field $\hat{v}(x,t)$ which is obtained by interpolation. It is assumed that both velocity fields v and $\hat{v}$ satisfy a Lipschitz condition, i.e. there exist positive Lipschitz constants L_v and $L_{\hat{v}}$ so that $\|v(x_1,t) - v(x_2,t)\| \leq L_v\|x_1 - x_2\|$ and $\|\hat{v}(x_1,t) - \hat{v}(x_2,t)\| \leq L_{\hat{v}}\|x_1 - x_2\|$ for all x_1, x_2 and t of the domain of v respective $\hat{v}$. Then the trajectories are given by the equations

$$\frac{dx}{dt} = v(x,t) \qquad \text{and} \qquad \frac{d\hat{x}}{dt} = \hat{v}(\hat{x},t) \quad . \tag{3}$$

If it is assumed that both trajectories have the same starting position $x(t_0) = \hat{x}(t_0) = x_0$, the difference between them at a later time t is:

$$\|x(t) - \hat{x}(t)\| = \left\| \int_0^t (v(x,s) - \hat{v}(\hat{x},s))\, ds \right\| \tag{4}$$

$$\leq (C_t h_t^2 + C_x h_x^2)t + L_{\hat{v}} \int_0^t \|x(s) - \hat{x}(s)\|\, ds \tag{5}$$

where C_t and C_x are constants. Now applying the Gronwall lemma (see [5]), the following estimate for the global discretization error caused by interpolating the velocity field is obtained:

$$\|x(t) - \hat{x}(t)\| \leq \frac{C_t h_t^2 + C_x h_x^2}{L_{\hat{v}}}(e^{L_{\hat{v}}t} - 1) \quad . \tag{6}$$

An estimation for the global error of the numerical integration of the exact velocity field v is given by

$$\|x(t) - x_\Delta(t)\| \leq \frac{C\tau^p}{L_\Delta}(e^{L_\Delta t} - 1) \tag{7}$$

where L_Δ is the Lipschitz constant of the integration scheme (compare [4, 5]).

If the constants C and L_Δ are comparable to $\max(C_t, C_x)$ and $L_{\hat{v}}$ respectively, and if the integration step τ is comparable to the time discretization step h_t and smaller than the cell size ($\|v\|\tau \leq h_x$), then equations (6) and (7) show that an integration scheme of order $p > 2$ should produce an error which is negligible compared to the discretization error. This means that an integration method of order $p \geq 2$ should be accurate enough with regard to the discretization error. To verify this assumption, we have implemented standard and extrapolated Runge-Kutta algorithms up to order 6 and extrapolated midpoint rules up to order 8, and tested these algorithms in several applications (compare Section 5). Furthermore, we have implemented embedded Runge-Kutta methods with error monitoring and adaptive step size control (see Section 4) to create a problem adapted grid Δ. Because of the adapted grid, these algorithms have a better performance and a higher accuracy than the simple Runge-Kutta methods.

3 Stiff Sets of Ordinary Differential Equations

In many practical applications the ordinary differential equations are *stiff*, for instance equations describing chemical reactions or the equations resulting from the semi-discretization of parabolic equations. Stiff systems were observed for the first time in 1952 by Curtis and Hirschfelder, when solving the problem of chemical reaction of a multicomponent system using explicit Runge-Kutta method. In this chemical experiment some components react in a very short time achieving the equilibrium state while the other components change very slowly.

It is important to distinguish between the stability of the ordinary differential equation and the stability of the integration method. In the case of the chemical reaction, explicit Runge-Kutta methods fail even for very small integration steps. On the other hand, the stability properties of the system are very good, i.e. the explicit Runge-Kutta integration method is unstable for stiff systems.

Stiff systems are in some sense characterized by the Jacobian $D_x v$ of v. The concept of stiffness can be mathematically defined but we concentrate on a characterization which is valuable in practical applications. We say that a system of ordinary differential equations is stiff in the *time* interval (t_0, t) if

$$(t - t_0)\|D_x v\| \gg 1 \quad \text{or equivalent} \quad (t - t_0)L \gg 1 \tag{8}$$

where L is the Lipschitz constant of v. If we use the $\mathcal{L}^2$ matrix norm, then $\|D_x v\|$ is equal to the absolute value of the largest eigenvalue of the matrix. In this section we consider fluid flow data and analyze under which circumstances the particle tracing equation (1) may show stiffness.

If a system of ordinary equations is stiff in some regions, then the corresponding linearized problem inherits this stiffness (compare [5]). Thus, only the linearized problem is considered. Furthermore, because the time dependent part is linear in t, it is not relevant for the considerations in this section. Thus, we only analyze the part of the ordinary differential equation that contains the Jacobian of the velocity:

$$\frac{dx}{dt} = D_x v \cdot x \quad . \tag{9}$$

In order to determine whether particle tracing for a given flow is stiff or not, the eigenvalues of the Jacobian of the velocity field have to be analyzed. The Jacobian can be decomposed into three parts: $D_x v = \Sigma + \Theta Id + \Omega$ where Σ represents the shear of the fluid flow and is a symmetric matrix, Θ is the expansion, Id the identity matrix, and Ω the rotation, which is an skew-symmetric matrix. The rotation matrix is of the form

$$\Omega = \frac{1}{2} \begin{pmatrix} 0 & -\omega_3 & \omega_2 \\ \omega_3 & 0 & -\omega_1 \\ -\omega_2 & \omega_1 & 0 \end{pmatrix} \tag{10}$$

where the vector ω is the rotation of the velocity, i.e. $\omega = \nabla \times v$.

If the shear of the flow field is large, e.g. near walls or within turbulence, the eigenvalues of Σ which correspond to the shear directions are large. It follows that the norm of the Jacobian is large and satisfies condition (8). In such situations it is recommended to switch to a special integration method, called implicit. We are coming back to this problem in the next section.

The norm of the matrix Ω can be easily computed due to its special form:

$$\|\Omega\| = \frac{1}{2}|\omega| = \frac{1}{2}\sqrt{\omega_1^2 + \omega_2^2 + \omega_3^2} \tag{11}$$

which is the vorticity of the flow. A typical example, where vorticity is relevant, is a circular flow. In this case particle tracing becomes a stiff problem. Additionally, such a circular flow shows also strong shear deformations.

4   Multi-Stage Methods

Due to the disadvantages of multi-step algorithms mentioned above, only multi-stage or single-step methods are considered. A detailed description of integration methods is found in [5, 9].

The well-known Runge-Kutta schemes (RKp) are the most famous explicit multi-stage algorithms. The letter p denotes the order of the integration scheme. The simplest one is Euler's method (RK1) and an example of RK2 is Heun's method. A different class of single-step algorithms are the extrapolation schemes, for instance the extrapolation of Euler's method (RK1Xp) or the extrapolation of the midpoint rule (MPXp). The order of RK1Xp rises by one after each extrapolation step and the order of MPXp increases by two after each step.

An improvement of explicit single-step algorithms are adaptive explicit single-step methods. The general idea of adaptive algorithms is to compute two trajectories with different integration schemes for each time step. If the difference between the endpoints of these traces is larger than a given tolerance, the computation is repeated with a smaller integration step size.

As adaptive schemes, embedded Runge-Kutta schemes are well suited. Here, the different integration methods are Runge-Kutta algorithms of different order and *embedded* means that the coefficients for the evaluation of the velocity field are all equal up to the lower order. Embedded Runge-Kutta methods are denoted RK$p(q)$ where p is the integration order and q is the order for the error estimation. We have implemented RK2(1), RK3(2) and RK4(3). Theses algorithms are represented with the help of the Butcher schemes, where the coefficients of the method can easily be seen (compare [2]):

$$
\begin{array}{c|cc}
0 & & \\
1 & 1 & \\
\hline
 & 1/2 & 1/2 \\
 & 1 & 0
\end{array}
\qquad
\begin{array}{c|ccc}
0 & & & \\
1 & 1 & & \\
1/2 & 1/4 & 1/4 & \\
\hline
 & 1/6 & 1/6 & 2/3 \\
 & 1/2 & 1/2 & 0
\end{array}
\qquad
\begin{array}{c|ccccc}
0 & & & & & \\
1/2 & 1/2 & & & & \\
1/2 & 0 & 1/2 & & & \\
1 & 0 & 0 & 1 & & \\
1 & 1/6 & 1/3 & 1/3 & 1/6 & \\
\hline
 & 1/6 & 1/3 & 1/3 & 1/6 & 0 \\
 & 1/6 & 1/3 & 1/3 & 0 & 1/6
\end{array}
$$

Adaptive Runge-Kutta algorithms RK2(1), RK3(2) and RK4(3).

Now we come back to stiff sets of ordinary differential equations. For a proper integration of such equations, linear implicit multi-stage methods are introduced. To see how implicit algorithms work, we consider equation (1) and write down the explicit Euler scheme for integrating this equation with step size τ:

$$x(t_n) = x(t_{n-1}) + \tau v(x(t_{n-1}), t_{n-1}) \quad . \tag{12}$$

The method is called explicit because the new value $x(t_n)$ is given explicitly in terms of the old value $x(t_{n-1})$. Whereas, the implicit Euler scheme is:

$$x(t_n) = x(t_{n-1}) + \tau v(x(t_n), t_n) \quad . \tag{13}$$

In general this is some nasty set of nonlinear equations that has to be solved iteratively at each step. In order to make implementation easier and to improve performance, the problem can be linearized. This leads to the linear implicit Euler method:

$$x(t_n) = x(t_{n-1}) + \tau \left(Id - \tau D_x v(x(t_{n-1}))\right)^{-1} \cdot v(x(t_{n-1}), t_{n-1}) \tag{14}$$

where $D_x v$ denotes the Jacobian of the vector field v at the point $x(t_{n-1})$. Since at each time step a matrix has to be inverted, linear implicit methods are slower than explicit ones. Therefore, linear implicit algorithms should be used only if the system of differential equations is stiff. Also remark that a linear implicit scheme converges to its corresponding explicit scheme for $\tau \to 0$.

We have implemented the linear implicit Euler method (LIRK1), the linear implicit midpoint rule (LIMP) and extrapolation schemes of both algorithms, abbreviated LIRK1Xp and LIMPXp respectively. Again the order of LIMPXp increases by two after each extrapolation step.

An improvement of implicit algorithms are again adaptive linear implicit multi-stage methods. Kaps and Rentrop showed in [6] that the smallest order for which embedded adaptive linear implicit Runge-Kutta methods are possible is order 4(3). We have implemented such an integration scheme and abbreviated it LIRK4(3).

We have implemented all integration algorithms mentioned in this section within the framework of the IRIS Explorer visualization environment.

5 Comparison of Integration Methods

In many applications vector fields arise that are very rough, so that higher order algorithms like extrapolation methods do not work well. Furthermore, due to the fact that the vector fields are strongly varying, the integration steps have to be very short. Thus, extrapolation schemes become very expensive. The explicit Euler method does not work well either. The first order integration method is not able to follow particles properly in turbulent flows.

Therefore we compare the performance of adaptive and non-adaptive Runge-Kutta methods only. The efficiency of the algorithms is measured by the number of evaluations of the velocity field. Also the number of calculated particles is listed in the following table. The evaluation of the velocity field corresponds to the point location and velocity interpolation steps of the particle tracing algorithm, which are the most time consuming operations [7]. For the test three data sets were used with different characteristics. The first one corresponds to the simulation of a vortex breakdown where the gradients of the velocity field are strongly varying. The second data set corresponds to the simulation of the air flow in a clean air room. The third data set is a laminar cross-cylinder flow. The velocity field of this data set is very smooth.

		RK2	RK3	RK4	RK2(1)	RK3(2)	RK4(3)
vortex	integr. steps	3694	1849	1233	691	239	166
	v-evaluations	7387	5545	4929	2582	1957	2211
air flow	integr. steps	5000	4147	3290	146	53	46
	v-evaluations	9999	12439	13157	636	225	235
laminar flow	integr. steps	291	146	97	94	34	29
	v-evaluations	581	436	387	346	180	189

In these tests the step sizes of all methods and the tolerances of the adaptive algorithms are chosen so that the resulting trajectories look identical and so that scaling down the step size of an integration scheme would not change its resulting trace. The speed differences are relative small in laminar flows, since all algorithms can integrate with a large step size. The speedup of adaptive methods is much greater in turbulent flow fields, because the non-adaptive methods have to use a small step size everywhere.

Furthermore, the tests confirm our consideration that RK3(2) is faster than RK4(3), if the integration is not more accurate than necessary. On the other hand RK2(1) is a lot slower than RK3(2). The scheme RK2(1) uses RK2 and Euler's method to determine the step size. Usually, the difference between the trajectory of the Euler scheme and that of the RK2 algorithm is relative large, and thus a relative small step size is chosen (compare Section 4). Hence, the RK2(1) method is significantly slower than the RK3(2) algorithm.

Now we compare the adaptive Runge-Kutta methods of our IRIS Explorer particle tracing module with the NAG-Advect-Simple module that uses an RK4(5) algorithm. The vortex data set is used to compare the computational time of the algorithms. The computational time of the NAG-Advect module is set to 100%.

NAG-Advect	RK4(3)	RK3(2)	RK2(1)
100%	86%	82%	109%

Relating to stiff problems, two example data sets where stiffness occurs are shown and described in Figure 1 and 2. These two figures are also displayed as colored images in the Appendix.

Figure 1 shows a flow in a floating-zone furnace for crystal growing. Because of the rotational symmetry, the stream bands should be closed. On the left hand side the stream bands are computed by the linear implicit Euler scheme and on the right hand side the classical Runge-Kutta method of order four is used. In this example both algorithms use the same step size and the same number of steps. On the left hand side the bands are closed but on the other side they are not. This is a typical data set where explicit methods fail and implicit algorithms have an advantage. The black planes define the starting positions for the particles.

Figure 2 shows a test data set of a cylindrical air flow caused by different temperatures on the walls of the box. Because of symmetry all trajectories should be closed. Here we have used both explicit and linear implicit RK4(3). At first glance both methods close their particle traces but after some 50 circulations the trajectories of the explicit Runge-Kutta scheme become thick (grey circles), whereas the traces of the implicit algorithm remain closed (black circles).

In both examples the particle tracing using the implicit algorithm took about twice as long as that using the explicit equivalent. Linear implicit integration methods are slower than their explicit counterparts but in stiff cases they are very helpful.

6   Conclusion

As a result of careful analysis of numerical efficiency and accuracy of different integration methods on discrete data, it is shown that an adaptive RK3(2)

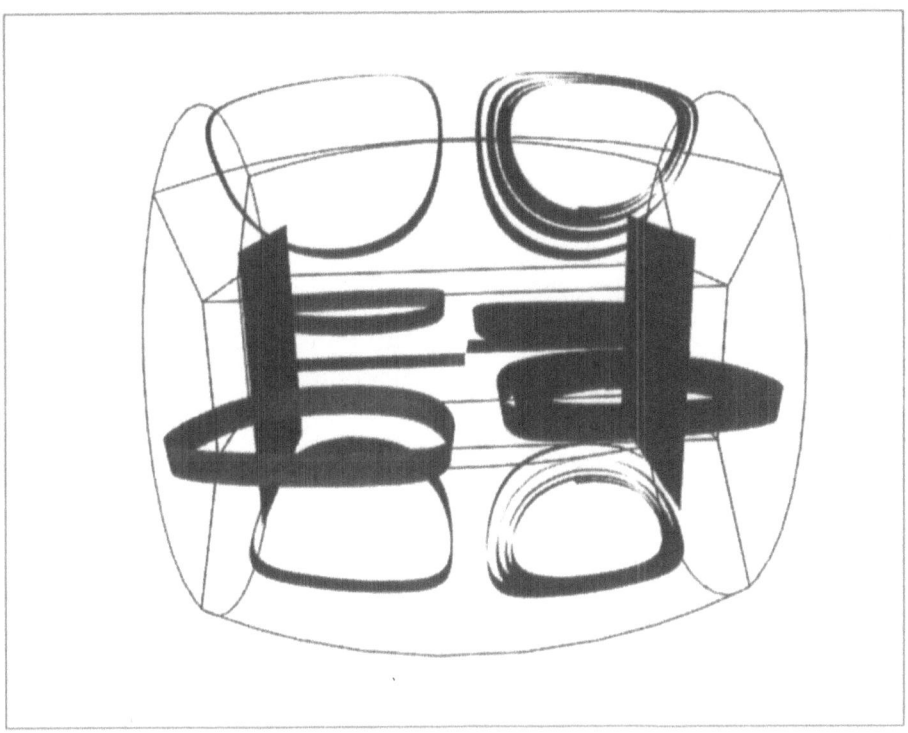

Fig. 1. Flow in a floating-zone furnace for crystal growing. On the left hand side the stream bands are computed by the linear implicit Euler scheme and on the right hand side the classical Runge-Kutta method of order four is used.

scheme is accurate enough in relation to the interpolation error and significantly more efficient than higher order integration algorithms. Furthermore, we have implemented implicit integration methods to handle stiff systems of ordinary differential equations. These algorithms are slower than explicit methods but in stiff data sets when explicit methods fail, they can create proper trajectories.

Based upon these results we have implemented a particle tracer as a module within the IRIS Explorer visualization environment.

Due to the fact that implicit integration is very expensive and that, in general, systems are stiff only in some localized regions, we intend to implement *partitioned* methods as a future extension. These are adaptive methods with a mechanism to switch between explicit and implicit integration depending on the eigenvalues of the Jacobian at the current point. If $\tau L \leq C$, the explicit integration is used else the implicit one. Here C is a method dependent constant (compare [3]) and L is the Lipschitz constant of the velocity field. If the currently used method is the implicit one, L can be approximated by the norm of the Jacobian (see Section 3) else L can be estimated by intermediate values of

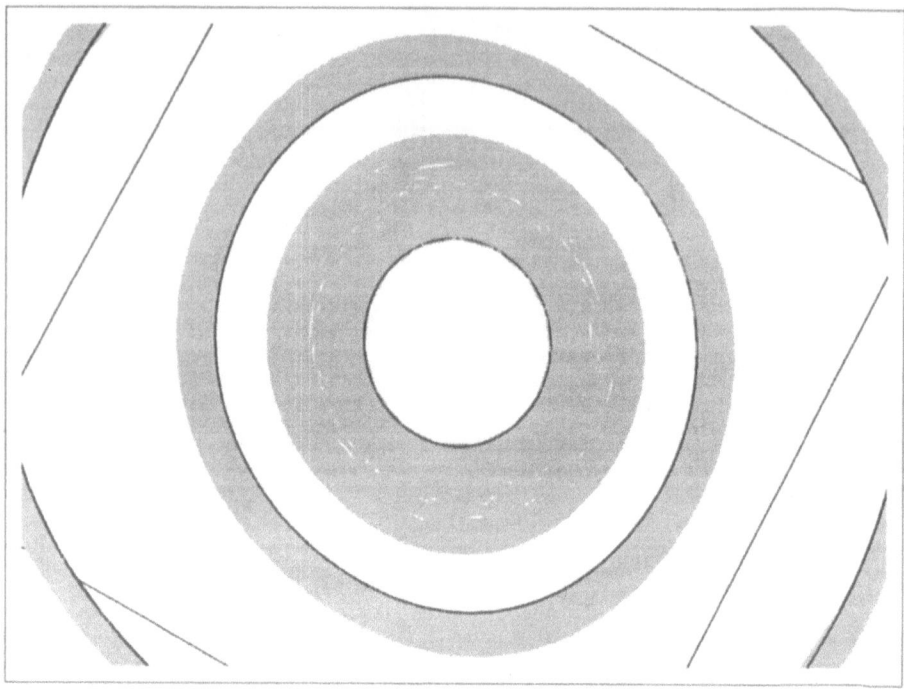

Fig. 2. Test data set of an cylindrical air flow caused by different temperatures on the walls of the box. Here we have used both explicit and linear implicit RK4(3). The trajectories of the implicit Runge-Kutta scheme remain closed (black circles), whereas the traces of the explicit method become thick (grey circles).

the explicit multi-stage method in the following way:

$$L \geq \frac{\|v(x^{(i)}(t_n)) - v(x^{(j)}(t_n))\|}{\|x^{(i)}(t_n) - x^{(j)}(t_n)\|} \quad . \tag{15}$$

7 Acknowledgments

We are grateful to H. Weimann from the Lehrstuhl für Werkstoffwissenschaften VI of the University of Erlangen for making the data set of a floating-zone furnace for crystal growing available to us. Also we would like to thank M. Breuer, S. Enger and K. Wechsler from the Lehrstuhl für Strömungsmechanik of the University of Erlangen for the other data sets and for their valuable remarks. Finally, we wish to thank the anonymous reviewers of this paper for their helpful comments.

References

1. B. Becker, D. A. Lane, and N. L. Max. Unsteady Flow Volumes. In G.M. Nielson and Silver D., editors, *Visualization '95*, pages 329–335, Los Alamitos, CA, 1995. IEEE Computer Society, IEEE Computer Society Press.
2. J. C. Butcher. Coefficients for the study of Runge-Kutta integration processes. *J. Austral. Math. Soc.*, 3:185–201, 1963.
3. J. C. Butcher. Order, stepsize and stiffness switching. *Computing*, 44:209–220, 1990.
4. D. L. Darmofal and R. Haimes. An Analysis of 3-D Particle Path Integration Algorithms. In *Proceedings of the 1995 AIAA CFD Meeting*, 1995.
5. P. Deuflhard and F. Bornemann. *Numerische Mathematik II: Integration gewöhnlicher Differentialgleichungen*. Walter de Gruyter, Berlin, New York, 1994.
6. P. Kaps and P. Rentrop. *Numerische Mathematik*, 33:55–68, 1979.
7. D. N. Kenwright and D. A. Lane. Optimization of Time-Dependent Particle Tracing Using Tetrahedral decomposition. In G. M. Nielson and Silver D., editors, *Visualization '95*, pages 321–328, Los Alamitos, CA, 1995. IEEE Computer Society, IEEE Computer Society Press.
8. D. N. Kenwright and G. D. Mallinson. A 3-D Streamline Tracking Algorithm Using Dual Stream Functions. In A. E. Kaufman and G. M. Nielson, editors, *Visualization '92*, pages 62–68, Los Alamitos, CA, October 1992. IEEE Computer Society, IEEE Computer Society Press.
9. William H. Press, Saul A. Teukolsky, William T. Vetterling, and Brian P. Flannery. *Numerical Recipes in C: The Art of Scientific Computing*. Cambridge University Press, Cambridge, New York, Victoria, second edition, 1992.
10. D. Stalling and H.-C. Hege. Fast and resolution independent line integral convolution. In *Computer Graphics Proceedings*, Annual Conference Series, pages 249–256, Los Angeles, California, August 1995. ACM SIGGRAPH, Addison-Wesley Publishing Company, Inc.

Editors' Note: see Appendix, p. 177 for colored figures of this paper

Creating Evenly-Spaced Streamlines
of Arbitrary Density*

Bruno Jobard and Wilfrid Lefer

Laboratoire d'Informatique du Littoral, Calais, France
{jobard,lefer}@lil.univ-littoral.fr

Abstract. This paper presents a new evenly-spaced streamlines placement algorithm to visualize 2D steady flows. The main technical contribution of this work is to propose a single method to compute a wide
variety of flow field images, ranging from texture-like to hand-drawing
styles. Indeed the control of the density of the field is very easy since the
user only needs to set the separating distance between adjacent streamlines, which is related to the overall density of the image. We show that
our method produces images of a quality at least as good as other methods but that it is computationally less expensive and offers a better
control on the rendering process.

Introduction

The problem of visualizing vector fields has been widely addressed in the past
years because it has numerous applications. The main issue is to visualize properly the direction and magnitude of the flow. Spatial resolution techniques such
as arrow plots, streamlines or particles traces suffer from their spatial resolution that limits drastically their usefulness, in particular in the presence of a
turbulent flow. Moreover the effectiveness of the streamline and particle methods depends critically on the placement of the forming seed points. Texture-like
methods, such as Spot Noise [10] and LIC [1] produce dense field images showing the flow features in fine-grain detail. Another issue is to compute sparse flow
fields, laying stress on the visual appearance of the field, which produces hand-
drawing style images. Recently, Turk and Banks presented an imaged-guided
streamline placement to compute hand-drawing style representations of a flow
field [9]. This paper presents another effective algorithm for the placement of
evenly-spaced streamlines. The main technical contribution of this work is to
propose a single method to compute a wide variety of flow field images, ranging
from texture-like to hand-drawing styles. Indeed the control of the density of the
field is very easy since the user only needs to set the separating distance between
adjacent streamlines, which is related to the overall density of the image.

* This work was supported by the council of the Nord-Pas-de-Calais region.

1 Related work

Visualizing a vector field in a general manner requires high spatial resolution techniques to properly render fine-grain details. Such methods generally yield dense field representations. However there are situations where a sparse density image is needed, by instance to produce an illustration similar to those used to enhance the purpose of text fields in a book. Methods proposed to visualize a flow field falls into these two categories: dense field representations and hand-drawing style.

The first method for representing a flow field with high spatial resolution has been proposed by van Wijk [10]. The *spot noise* method creates a directional texture by superimposing many flow-oriented ellipses. Each ellipse is generated by projecting a spherical spot onto a surface and by advecting the spot with the direction and magnitude of the vector field at the projection point. This amounts to the flow field-controlled generation of a band-limited noise. Initially straight, the spots are now bent along short streamlines to follow the curvature of the vector field [2]. An important feature of this method is the local control on the generated image. More spots gives rise to more accuracy. With spot noise the generation time depends on the number of spots used to generate the texture. Consequently, by setting the number of spots a trade-off can be obtained between image quality and rendering time.

Another interesting method is the *line integral convolution* (LIC) proposed by Cabral and Leedom [1] [4] [8]. A LIC texture is generated by convoluting an input texture with a streamline-oriented one dimensional filter kernel. The images obtained with this technique are very effectives, showing more details than the previously enumerated one. But this is obtained at the expense of computation. This computation cost is mainly due to the number of streamlines that have to be computed, and let us notice that even with the *fastLIC* method [8], several streamlines cover each single pixel of the resulting texture, giving rise to frequent recomputations.

The image-guided streamline placement method proposed in [9] uses a stochastic mechanism to iteratly refine the placement of the streamlines. First an initial set of randomly placed streamlines is created. Then this set of streamlines is updated using three valid operations: (1) changing the position and/or length of a streamline, (2) joining streamlines that nearly abut, and (3) creating a new streamline to fill a gap. At each step of the refinement process a small change, i.e. a combination of the three operations mentioned above, is randomly performed. An energy function consults a low-pass-filtered (blurred) version of the image to measure the variation of energy between the current and the updated images and the modification is only accepted if the variation of energy if negative. This method produces high quality images but the convergence is very slow and obtaining a good visual appearance often requires several minutes for each image to be computed. Moreover this method is not suitable for dense field images because of the combinatorial explosion of the possible modifications at each step.

The remainder of this paper is organized as follows. In section 2 we present a method for effective user-controlled evenly-spaced streamlines placement. Section 3 describes the use of the method to produce hand-drawing style images and we compare our approach with the image-guided streamline placement from Turk and Banks. In section 4 we show how texture-like images can be obtained and we discuss the advantages and drawbacks of this method compared to LIC. We conclude and offer directions for future research in section 5.

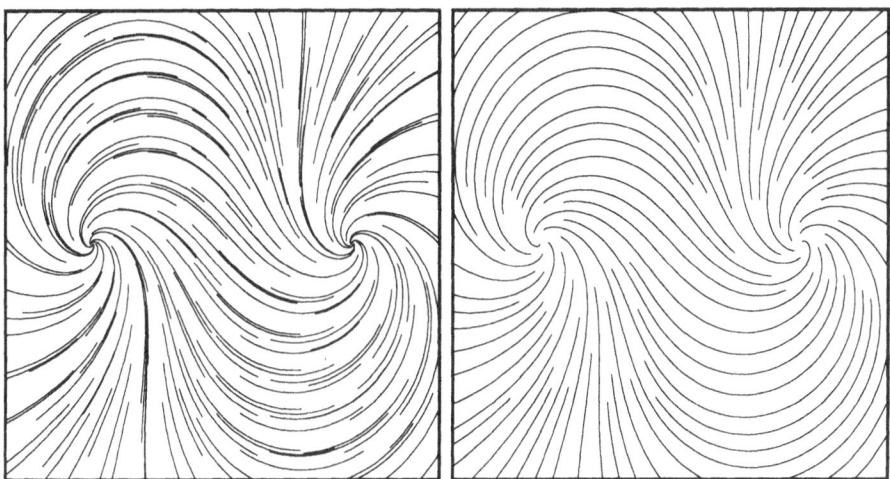

Fig. 1. (a) Long streamlines with seed points placed on a regular grid (left); (b) Same flow field computed using our streamline placement method (right)

2 Algorithm overview

The goal of this work is to produce long and evenly spaced streamlines in a single pass. The basic principle of our algorithm is similar to a method presented by Max et al. in [6]. The goal of their work was to cover a 3D surface (not necessary tangential to the field) with a set of streamlines. Once a seed point has been selected in the field, they make a streamline growing beyond that point backward and forward. The growing process is stopped when the streamline reaches an edge of the surface, a singularity in the field (source or sink) or becomes too close to another streamline. The streamline is then divided into a set of small segments of contrasting color and projected onto the surface. Although Max's method was intended to visualize a flow on a 3D surface, it can be generalized to all kinds of steady 2D field.

We extend this work in the following manner. First we give a number of precisions concerning the implementation of the algorithm together with a couple of optimizations. Second we show how the algorithm can be controlled by the user to produce a wide range of flow fields images, ranging from hand-drawing to LIC-like style.

An important feature of the algorithm is that it processes in a single pass (as compared to Turk's progressive refinement approach). To compute an image, a number of streamlines are calculated until a user-fixed density level has been obtained. Computing a new streamline is achieved in the following manner. A new seed point is chosen at a minimal distance apart from all existing streamlines. Then a new streamline is integrated beyond the seed point backward and forward until either it goes too close from another streamlines or it leaves the 2D domain over which the computation takes place. The algorithm ends when no more valid seed point can be found. Figure 1b shows an image obtained with our algorithm and figure 1a shows the same flow field visualized using a distribution of the seed points over a regular grid. The three following sections detail three important points of the algorithm: the control of the distance between adjacent streamlines, seed points selection and streamline integration.

2.1 Control of the separating distance

Density is a global feature of the field. However we need to express it as a local feature in order to have a local control on the texture generation. We express the density as the distance between two adjacent streamlines. Let d_{sep} be this distance. Hence the control of the density of the field is achieved by controlling that there is not any pair of streamlines apart from a distance lower than d_{sep}. This control occurs during the construction of each streamline. During the construction, a new sample point is valid only if it is at a separating distance greater than d_{sep}. If it is not the case, the streamline is stopped in this direction (during construction streamlines grow in both directions independently). In order to make the computation of the separating distance faster, rather than computing the exact distance from the new seed point location to the streamline, we compute the distance from the seed point to the sample points along the streamline. To make this approximation acceptable, the sample points on a streamline must be evenly spaced and the distance between them must be smaller than d_{sep}. Thus, a new sample point is valid if the distance between it and all the existing sample points is greater than the separating distance. Since this test has to be computed for all the sample points, it must be as fast as possible. To accelerate the computation of the distance, we use a cartesian grid superposed to the vector field domain, the width and height of a cell being exactly d_{sep}. Each cell contains a list of pointers to the sample points located within the cell. Thus, given a new seed point location, the cell containing the location is easily determined. Let us call this cell the local cell. The distance has to be computed only for the sample points located either within the local cell or within the eight cells surrounding the local cell. In practice, we have noticed that an average of 5-7 distance computations is necessary to determine if a new sample point is valid or not. The denser is the grid, the less comparisons are required.

remark: Practically, we consider two important distances, d_{sep} and d_{test}. d_{sep} is the separating distance given by the user. It represents the minimal distance between seed points and streamlines. d_{test} is a percentage of d_{sep}. It corresponds to the minimal distance under which the integration of the streamline will be stopped in the current direction. We found $d_{test} = 0.5 \times d_{sep}$ gives good visual result by producing long streamlines. For instance, images of Figure 2 have been calculated with two different values of d_{test}.

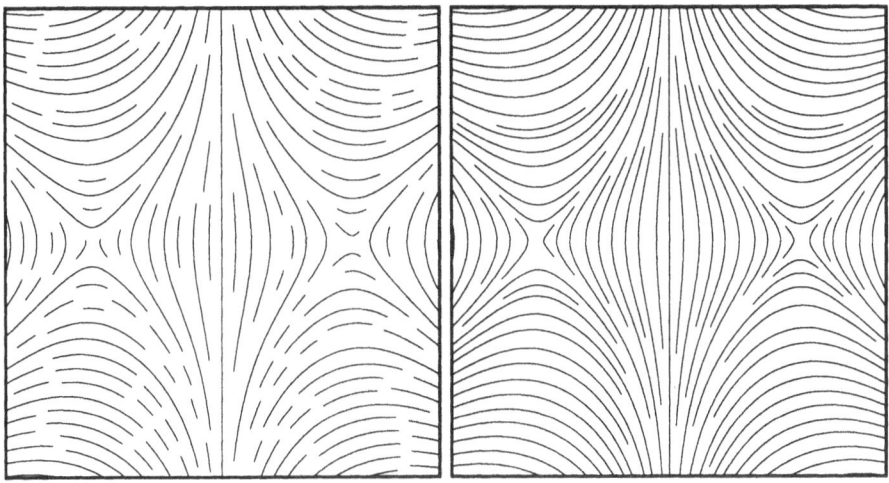

Fig. 2. increasing difference between d_{sep} and d_{test} lengthen streamlines; (left) $d_{1_{test}} = 0.9 \times d_{sep}$; (right) $d_{2_{test}} = 0.5 \times d_{sep}$

2.2 Seed points selection

When using streamlines for vector field visualization, a common problem is to select proper seed points for path tracking. Dovey proposed two approaches to resample non-uniformly spaced grids in order to achieve an uniform density of vector glyphs [3]. A vector field is represented with short segments oriented by the flow. In case of short streamlines or hedgehogs, the resulting image mainly depends on the distribution and density of the seed points over the domain. In case of long streamlines a constant density of seed points do not ensure a good distribution of the streamlines.

In order to obtain a good visual appearance of the flow field, an accurate seed point selection has to be perform. The principle of our algorithm is to derive all the seed points possible to find from an existing streamline before trying with another existing one. The proposed seed points are chosen at a distance $d = d_{sep}$ from the sample points of each streamline. Our algorithm uses a queue to store the newly created streamlines which are processed from the older one to the more recently created one. The algorithm is given below.

```
Compute an initial streamline and put it into the queue
Let this initial streamline be the current streamline
Finished := False
Repeat
    Repeat
        Select a candidate seedpoint at d = dsep apart from the current streamline
    Until the candidate is valid or there is no more available candidate
    If a valid candidate has been selected Then
        Compute a new streamline and put it into the queue
    Else
        If there is no more available streamline in the queue Then
            Finished := True
        Else
            Let the next streamline in the queue be the current streamline
        Endlf
    Endlf
Until Finished=True
```

Figure 3 shows all the streamlines the algorithm has been able to derive from the first streamline for a given vector field.

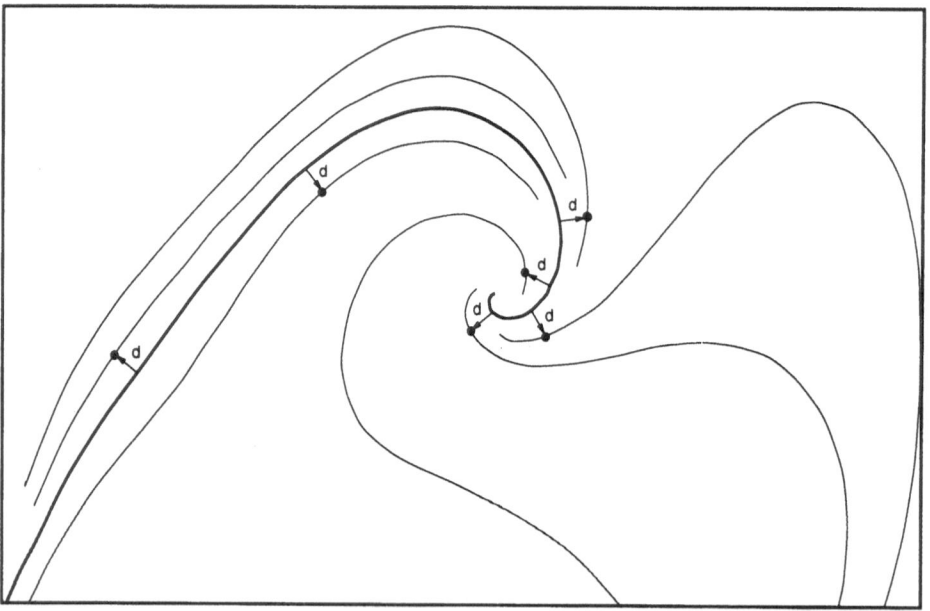

Fig. 3. streamlines are derived from the first (thick) one by choosing seed points (circles) at a distance d = d_{sep} from it

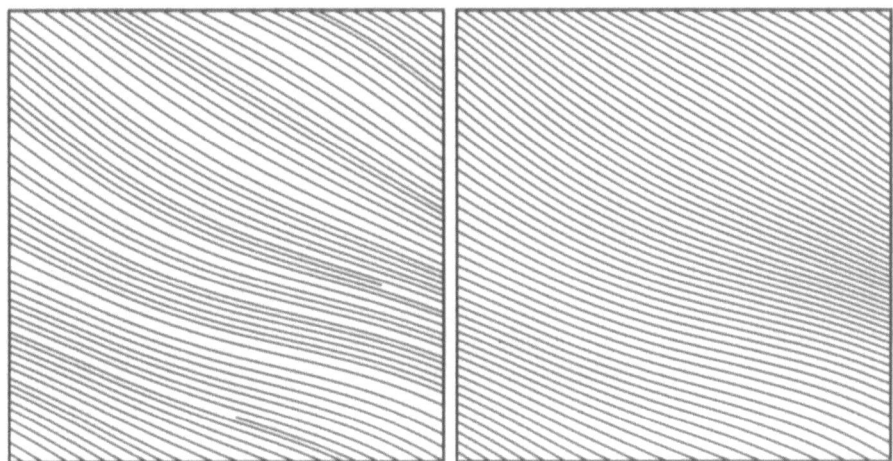

Fig. 4. Two seed points selection method with the same density of streamlines: (left) random selection, (right) our selection method

For sparse illustrations choosing to start streamlines close to existing ones gives better visual results than selecting seed points in a random fashion (see figure 4) but is more time-consuming. For dense texture generation, the quality of images produced with various seed point selection methods is quite similar.

2.3 Streamline integration

In order to measure a consistent separating distance between a point and a streamline, sample points along a streamline must be evenly spaced (see section 2.1). Many integrators are able to produce such evenly spaced sequences of sample points. They can be classified into three categories:

- fixed step size integrators such as Euler, Midpoint or Runge-Kutta methods with normalized vector fields,
- non constant or adaptive step size integrators with a post interpolation phase such as cubic Hermite-interpolation, which deal with large distances between neighboring sample points and curvature of the streamlines [8],
- continuous integration methods such as DOPRI5, which is a fifth order Runge-Kutta integrator with adaptive step size monitoring and fourth order error estimation and produces a dense output directly by using informations gathered at each step of the integration [5].

At present we use the Midpoint integrator but future investigations will concern the choice of a better integration method. In particular using an adaptive step size integrator will decrease the number of integrations required, reducing overall computation time.

Fig. 5. Image comparisons for separating distances of 6%, 3% and 1.5% of image width; left column: Image-guided placement; right column: Our streamline placement.

3 Hand-drawing style

Sparse illustrations of flows fields are the more interesting application of our method. Turk et al. proposed a method to compute such a representation in [9]. The method computes an initial set of streamlines which is then iterately refined until the global energy of the image falls below a fixed threshold or the user stops the process. The images obtained with this method are of great quality but the convergence of the iterative process is very slow and becomes much slower when desired density increases. Moreover the energy function used to measure the quality of the result is not directly related to the visual appearance of the image, requiring the appreciation of the user to stop the process.

The advantage of our method as compared to Turk's one is that it processes in a single pass, computing the final image directly. Figure 5 shows the same flow field computed by Turk's method with various refinement degrees and by our method and table 1 gives computation times necessary to produce the different images. We see that our method produces images of the same quality as Turk's ones but is less time-consuming.

separating distance	Image-guided placement	Our placement algorithm
6% image width	Fig 5a: stopped at 2 mn	Fig 5b: 4 seconds
3% image width	Fig 5c: stopped at 4 mn	Fig 5d: 9 seconds
1.5% image width	Fig 5e: stopped at 10 mn	Fig 5f: 17 seconds

Table 1. Computation times on a MIPS R4600 Processor, 132Mhz with 32Mo. Image-guided placement images and computation times were obtained with the Greg Turk's original publicly available algorithm.

Fig. 6. Hand-drawing style images computed without and with the tapering effect

Tapering effect. As stated in section 2.1 the actual distance between streamlines is not constant. Since the density is related to the distance between streamlines, disparities of density may appear in the resulting image. To reduce this visual artifact, Turk suggested to *taper* the ends of the streamlines by decreasing the thickness of the lines as they go closer to another one. In case of Turk's algorithm, this is achieved in a post-processing step. Our implementation allows to directly include the tapering optimization during the growing process of each streamline. The width of the streamline is computed using the following formula:

$$thicknessCoef = \begin{cases} 1.0 & \forall d \geq d_{sep} \\ \dfrac{d - d_{test}}{d_{sep} - d_{test}} & \forall d < d_{sep} \end{cases} \; ; \; thicknessCoef \in [0;1]$$

where d is the distance to the closer streamline (see section 2.1 for the definition of the different distances).
Fig 6 shows the same image computed without and with the tapering effect.

Glyph mapping. Once the streamline placement has been computed, the streamlines can be viewed as *skeletons* on which directional glyphs can be mapped. Figure 7 shows an image obtained by mapping such kind of icons onto the computed streamlines. This enables to add a directional information in the field.

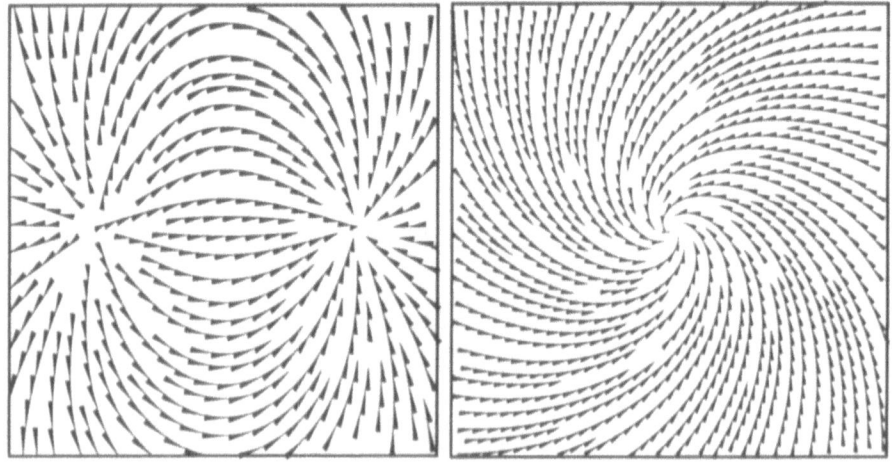

Fig. 7. examples of illustrations with glyph mapping

4 Texture generation

By decreasing the separating distance d_{sep}, the coverage of the streamlines becomes dense over the field. To depict the tangential component of the flow in a dense representation we have to differentiate close streamlines to be able to follow them over the field. This is achieved by mapping a periodic intensity function onto the streamline. Let us consider $f(x)$ a function which associates an intensity to every sample point on a streamline where x is the rank of the sample point within the streamline. f will give the shape of the intensity wave on the streamline. For instance, one may associate the two functions f_1 and f_2 given below:

$$f_1(x) = \frac{1}{2}(1 + sin(\frac{2\pi x}{N})) \quad and \quad f_2(x) = \frac{x \bmod N}{N-1}$$

where N is the length of the period as a number of sample points.

f_1 will give a smooth continuously increasing and decreasing intensity while the modulo function f_2 will produce discontinuous segments of increasing intensity in order to remove ambiguities about the orientation of the flow. Figures 8 and 9 show an image obtained using this pair of functions.

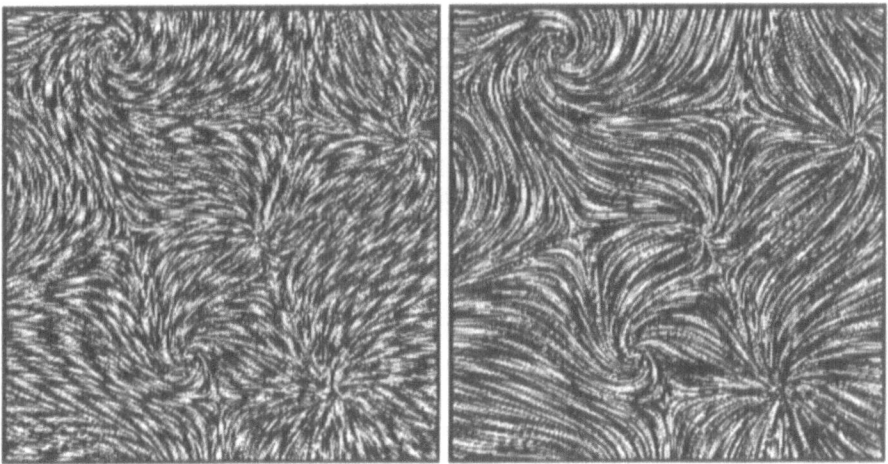

Fig. 8. textures generated with the f_1 function for intensity effects; (left) short and (right) long period of f_1

The images obtained with our method look like LIC images. In fact our algorithm is somehow a dual version of LIC. With LIC, for each pixel p of the output image, one integrates a particle path centered on p forward and backward and then averages intensities of the input texture pixels to get the intensity of the pixel being calculated. In our method, we find an optimal dense coverage of the field which minimizes the number of integrated streamlines and then associate an intensity to each pixel of all the streamlines.

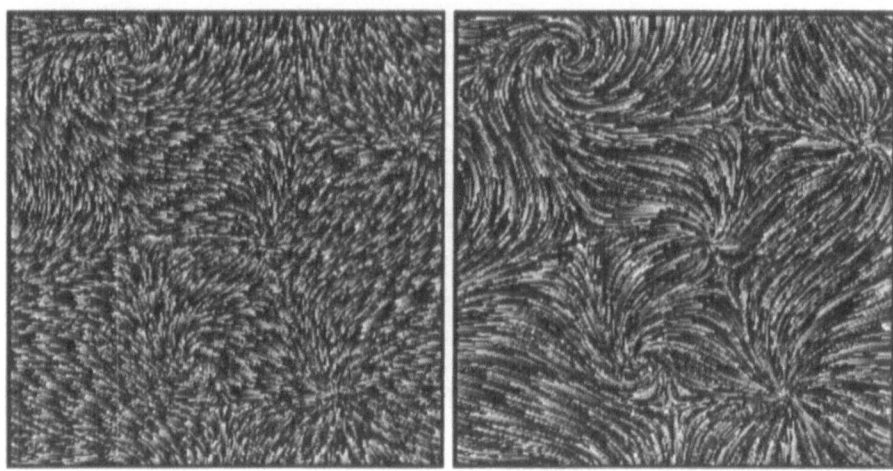

Fig. 9. textures generated with the f_2 function for intensity effects; (left) short and (right) long period of f_2

Now let us point on the advantages of our method over LIC. The first advantage is that our method does not require an input texture to process. With LIC a problem arises when one want to change the length of the apparent streamlines. In case of LIC, it is necessary to change the length of the convolution filter, which gives rise to a computation overhead. It is stated in [1] that doubling the length increases the computation time by a factor of four. Okada and Lane have proposed another solution, which consists in executing the LIC algorithm twice, the resulting image of the first execution being used as the input of the second execution [7]. This method results in lines which are more easily distinguishable but it concentrates the pixels values in a narrow range of intensity, which decreases the global contrast. This effect can be removed by applying post-filters to the final image, but this is done at the expense of a computation overhead (approximately by a factor of two).
With our approach we are able to change the length of the apparent streamlines on demand by simply changing the length of the period of the function as described above, without recomputation of any streamline.

In the traditional LIC algorithm, a streamline integration and a line convolution are computed for each individual pixels, so most of the time is spent in convolution and integration operations. Decreasing the number of streamlines computed would greatly benefit the LIC algorithm. Stalling and Hege proposed the *FastLIC* algorithm which reduces the overall number of streamline computations by sharing the line integral convolution information of each streamline with all the pixels it goes through [8].

With our approach, we compute the minimum number of streamlines necessary to cover entirely the 2D area over which the field is studied.

For instance, the computation of textures images of size 512×512 (such as figures 8 and 9) takes about 25 seconds on a R5000SC-64Mo based system. These results have been obtained using a random seed points selection method and a separating distance of 0.3%. As far as dense texture images are concerned, a drawback of our method is the aliasing effect due to the drawing of adjacent line segments of different colors. To remove this artifact we can smooth the final image by simply applying a blur filter.

5 Conclusion

We have presented an effective method to place long evenly-spaced streamlines with an accurate control on the density of the final image. By changing the separating distance between streamlines we are able to produce from sparse to dense representations of flow fields. We show that our method produces images of a quality at least as good as other methods but that it is computationally less expensive and offers a better control on the rendering process. Future investigations will concern a more efficient integrator, generalization to unsteady flows and real time animation.

References

1. Brian Cabral and L. Leedom. Imaging vector fields using line integral convolution. *Computer Graphics*, 27:263–272, jul 1993.
2. W. C. de Leeuv and Jarke J. van Wijk. Enhanced spot noise for vector field visualization. In *Proc. of Visualization '95*, pages 233–239. IEEE Press, Los Alamitos, CA, oct 1995.
3. Don Dovey. Vector plots for irregular grids. In *Proc. of Visualization '95*, pages 233–239. IEEE Press, Los Alamitos, CA, oct 1995.
4. Lisa K. Forsell. Visualizing flow over curvilinear grid surfaces using line integral convolution. In *Proc. of Visualization '94*, pages 240–247. IEEE Press, Los Alamitos, CA, oct 1994.
5. E. Hairer, S. P. Nørsett, and G. Wanner. *Solving Ordinary Differential Equations I - Nonstiff Problems*. Springer-Verlag, 1993.
6. Nelson Max, Roger Crawfils, and Charles Grant. Visualizing 3D velocity fields near contour surfaces. In *Proc. of Visualization '94*, pages 248–255. IEEE Press, Los Alamitos, CA, oct 1994.
7. A. Okada and D. Lane. Enhanced line integral convolution with flow feature detection. Technical Report NAS-96-007, NAS, jun 1996.
8. D. Stalling and H-C. Hege. Fast and resolution independent line integral convolution. *Computer Graphics*, 29:249–256, jul 1995.
9. Greg Turk and David Banks. Image-guided streamline placement. *Computer Graphics*, 30:453–460, jul 1996.
10. Jarke J. van Wijk. Spot noise: Texture synthesis for data visualization. *Computer Graphics*, 25(4):309–318, jul 1991.

Line Integral Convolution for 3D Surfaces

Xiaoyang Mao, Makoto Kikukawa, Noboru Fujita and Atsumi Imamiya

Department of Electrical Engineering and Computer Science
Yamanashi University, Japan

Abstract. *Line Integral Convolution(LIC) is a very powerful vector field visualization technique as it can effectively reveal the global and complex structures of a flow field. All the existing LIC algorithms, however, requires the one-to-one correspondence between input image pixels and grid cells, and hence restrict their use only for 2D/3D structured grids. In this paper, we present a new algorithm for convolving solid white noise on triangle meshes in 3D space, and extend LIC for visualizing the vector field on any arbitrary 3D surfaces, such as a contour surface output from the Marching Cube algorithm, or a surface of a 3D object represented implicitly by a part of a curvilinear or an unstructured grid.*

1 Introduction

Line Integral Convolution(LIC), which was first proposed by Cabral and Leedom[1] in 1993, has been attracting large attentions in the area of vector field visualization. By convolving white noise input images with one dimensional filter kernels along the local streamlines of vector fields, LIC can generate texture images intuitively and effectively showing the global property, as well as the details of local structures of flows.

Large vector fields, especially those fields representing complex flows like turbulence, are usually difficult to visualize through traditional visualization techniques. The techniques based on inserting geometrical primitives such as arrows, streamlines and particles into a flow field may either result in cluttering images or fail to capture some important features of flows due to inadequate sampling of vector fields. The texture based approach, however, is particularly useful in visualizing such kind of vector fields as it can generate an image which is a dense, but legible representation of a whole vector field. The first attempt to employ texture synthesis technique for vector field visualization was made by van Wijk [13]. He proposed a way to generate various textures (called *spot noise*) through convolving an input white noise image with randomly distributed 2D filter kernels(called *spot*). By choosing the shape of filter kernels to be elliptical and aligning the major axes of the ellipses with the directions of vectors, textures effectively depicting the global appearance of flows can be obtained. The spot noise technique has been further enhanced to deal with highly curved flows by deforming the shape of spots with local stream surfaces[3]. Instead of convolving with 2D filter kernels, LIC uses one dimensional filter kernels defined on local streamlines, and thus is easier to be implemented and can clearly visualize the details of small structures, such as the center of a vortex. Stalling and Hege [12] succeeded in a fast implementation of LIC. Their algorithm also allows output image resolutions to be chosen independent of vector field sizes, and hence the close up views of interested areas become available.

The purpose of this paper is to extend LIC for visualizing the vector field on arbitrary 3D surfaces. The application of such a technique is myriad. For example, in global environment simulation, it is usually natural to visualize the wind velocities as along the contour surfaces of air pressure. In aerodynamics simulation, the air flow over the surface of a space shuttle or aircraft is often of largest interest to scientists and engineers. Max, Crawfis and Grant have discussed several ways for visualizing the vectors near contour surfaces [6]. van Wijk [3, 13] and Forssell, et al. [2] have extended spot noise and LIC, respectively, for visualizing the flow on a 3D parametric surface represented as a 2D or a slice of 3D curvilinear grid. Their algorithms are realized as two mappings between the computational space and the physical space of a vector field. First the vectors in the 3D physical space are mapped into the computational space, which is a regular 2D Cartesian grid of unit cell size. Then the conventional spot noise or LIC algorithm is performed there, and finally the resulting 2D LIC image is mapped back onto the 3D surface in the physical space. By taking the advantage of texture mapping hardware, the resulting LIC texture mapped 3D surface can be rendered in realtime. Their approaches, however, may suffer from following problems inherently caused by the mapping between computational space and physical space:

1. Vectors might be inaccurately mapped into computational space due to the discrete approximation of the Jacobian matrices [11].
2. As a curvilinear grid usually consists of cells of drastically different sizes, the resulting LIC texture may be distorted after being mapped to the 3D surfaces. Local texture distortions can also be caused by a non-linear mapping between the computational space and the physical space.

For the second problem, Leeum and Forssell have suggested adjusting the sizes of filter kernels in the computational space according to the local cell sizes[2, 13]. This usually can not always resolve the problem thoroughly because the granularity of noise is also stretched or condensed with the mapping. However, adapting the granularity of noise in an input image to the local cell sizes in a vector field seems to be a very difficult task. Another disadvantage of their approaches is that their use has been restricted to structured grids consisting of quadrilateral cells.

Shen et al. [10] proposed an algorithm for visualizing the flow near a contour surface by volume rendering a 3D LIC image. As a flow is visualized through propagating in a 3D LIC image the opacity used for volume rendering, the global features of a flow can not be depicted with a single static image without animation. Also, their algorithm can only be applied to regular structured 3D grids.

The new algorithm presented in this paper solves these problems through the using of solid texturing [7, 8]. LIC is directly performed in 3D physical space on procedurally generated 3D solid white noise images[4, 8]. Here we assume the 3D surfaces is represented as triangle meshes, and vectors together with other scalar values are all stored at the vertices of triangles. This makes our algorithm very general, able to handle any kind of 3D surface, such as an isosurface generated with Marching Cubes algorithm [5], a parametric surface represented as a slice of a 3D curvilinear grid or the exterior of a 3D unstructured grid. In the latter case, the grid can be easily converted into a triangle meshes. The new algorithm has been implemented and experimentally applied to the visualization of supersonic flows over a spiked blunt body.

In the rest of the paper, we first give a brief review of LIC algorithm in Section 2. Section 3 outlines our new algorithm and discusses how to generate 3D solid white noise. The algorithm for convolving the solid noise image along the local streamlines in 3D physical space is given in Section 4. Section 5 addresses implementation issues and demonstrates some application results of the presented technique. Finally we conclude the paper by showing some future research directions.

2 Background

As the preliminaries to the description of the new algorithm in succeeding sections, here we give a overview of the original LIC algorithm presented by Cabral, et al. [1].

Given a vector field represented as a regular Cartesian grid, LIC algorithm takes as input a white noise image of the same size as the vector field, and convolves the image at each pixel with a 1D filter kernel defined along the local streamline in the vector field. As shown in Fig. 1, for a pixel $P(x, y)$, the local streamline in the vector field is calculated by integrating forward and backward cell by cell along local vector directions.

The forward integration is performed in the following way:

$$P_0 = (x + 0.5, y + 0.5),$$

$$P_i = P_{i-1} + \frac{V_{P_{i-1}}}{|V_{P_{i-1}}|} \Delta s_{i-1}.$$

$$\Delta s_{i-1} = min(D(P_{i-1}, P_{rl}), D(P_{i-1}, P_{tb}))$$

$$P_{rl} = \begin{cases} P_{right} & V^x_{\lfloor P_{i-1} \rfloor} > 0 \\ P_{left} & V^x_{\lfloor P_{i-1} \rfloor} < 0 \end{cases}$$

$$P_{tb} = \begin{cases} P_{top} & V^y_{\lfloor P_{i-1} \rfloor} > 0 \\ P_{bottom} & V^y_{\lfloor P_{i-1} \rfloor} < 0 \end{cases}$$

Where $V_{\lfloor P_{i-1} \rfloor}$ is the vector at the grid point $(\lfloor P^x_{i-1} \rfloor, \lfloor P^y_{i-1} \rfloor)$ and $V^x_{\lfloor P_{i-1} \rfloor}$, $V^y_{\lfloor P_{i-1} \rfloor}$ are the X, Y components of the vector $V_{\lfloor P_{i-1} \rfloor}$. $P_{right}, P_{left}, P_{top}, P_{bottom}$ are the intersections between the line passing through the point P_{i-1} along the

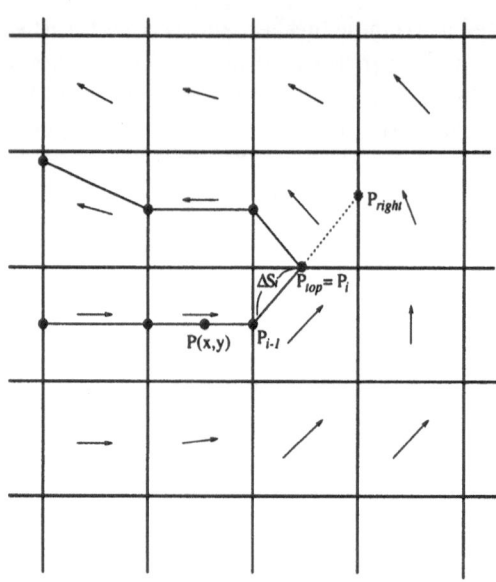

Fig. 1. Local streamline for a point $P(x, y)$ in a 2D vector field.

direction of $V_{\lfloor P_{i-1} \rfloor}$ and the four edges of the cell $(\lfloor P_{i-1}^x \rfloor, \lfloor P_{i-1}^y \rfloor)$, respectively. $D(P_{i-1}, P_c)(c : right, left, top, bottom)$ are the distances from P_{i-1} to the intersections. If a zero velocity or a vector pointing back to the cell to be left is encountered, then the integration terminates at that point before reaching the specified streamline length.

The backward integration P'_i is calculated simply by taking the opposite direction of the vector at each point. Now the output pixel value at the pixel $P(x, y)$ can be represented as

$$F_{out}(P) = \frac{\sum_{i=0}^{l} F_{in}(\lfloor P_i \rfloor) h_i + \sum_{i=0}^{l'} F_{in}(\lfloor P'_i \rfloor) h'_i}{\sum_{i=0}^{l} h_i + \sum_{i=0}^{l'} h'_i}$$

$$h_i = \int_{s_i}^{s_i + \Delta s_i} k(w) dw$$

where

$s_0 = 0, s_i = s_{i-1} + \Delta s_i$.
$k(w)$: the convolution filter .
$F_{in}(\lfloor P_i \rfloor)$: the input pixel value at pixel $(\lfloor P_i^x \rfloor, \lfloor P_i^x \rfloor)$.
l, l': the length of streamline in unit pixels at each side of the point.

Note that the local streamline generated here only depends on the direction of vectors but ignore the magnitude. More accurate streamline calculation using numerical integration was given in [12].

3   Solid Texuring for LIC

As mentioned in Section 1, it is difficult for the existing texture mapping based 3D LIC algorithm [2, 3] to generate LIC textures without distortion on 3D surfaces in case the underlying curvilinear grid consists of cells of different sizes or the mapping between computational space and physical space is non-linear. A distorted LIC texture may easily cause wrong conclusions about a flow. For example, with the LIC image on the space shuttle shown in [2], the flow over the wing area, where the granularity of input noise has been stretched due to the relative large cell sizes, tends to look faster than that at the head of the shuttle. To overcome such problem and also to get rid of the restriction to structured curvilinear grid, we present here a new approach utilizing the technique of solid texturing [7, 8].

As depicted in Fig. 2, to generate the 3D LIC texture on an arbitrary 3D surface, we shoot rays from the viewpoint through screen pixels into the 3D surfaces. Convolution of a 3D white noise image with filter kernels defined along the local streamlines are performed only at the visible ray-surface intersections.

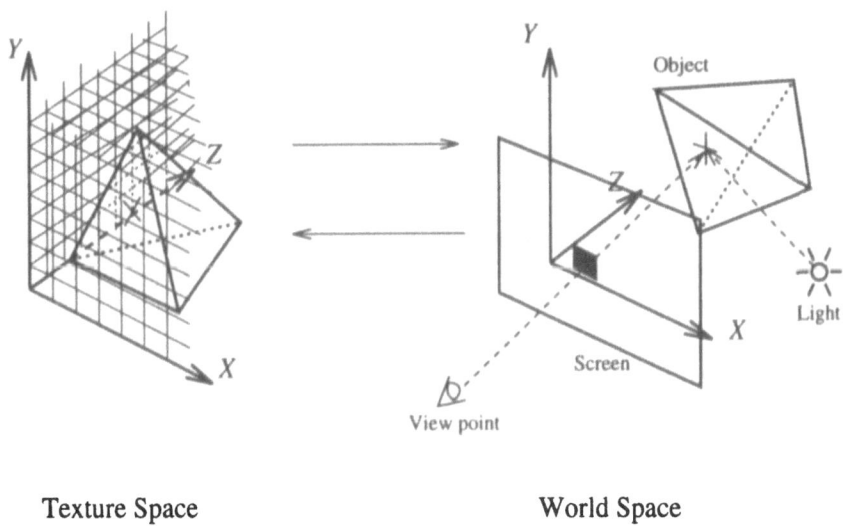

Texture Space   World Space

Fig. 2. LIC for 3D surfaces through solid texturing.

The solid white noise value at a point $P(x, y, z)$ in 3D space is evaluated procedurally with the method introduced in [4, 8]. Imagine a 3D *integer lattice* constructed by all the points having integer X, Y, Z coordinates is imposed onto the 3D surface. Each lattice point is assumed to be associated with a noise value $V(i, j, k)$ through a hash-like pseudo-random function of its coordinates i, j, k.

Our implementation uses a simple hashing function similar to that described in [4];

```
int Noise(int i,j,k)
{
  int n = INDEX[(a*i)&0x000000ff] + INDEX[(b*j)&0x000000ff]
          + INDEX[(c*k)&0x000000ff] ;
  return (WHITE_NOISE[n%TABLEN]);
}
```

Here, WHITE_NOISE is a table of randomly generated binary noise values and TABLEN is the size of the TABLE. INDEX contains pseudo-random integer values as the indirectional table of indices to the table WHITE_NOISE. As what we need is discontinuous white noise, for an arbitrary point on the 3D surfaces, its noise value can be obtained simply by taking the floor of coordinate values before using the hashing function.

$$V(x, y, z) = Noise(\lfloor x \rfloor, \lfloor y \rfloor, \lfloor z \rfloor)$$

Since it is necessary to have the texture fixed on the 3D surface independent of the position and orientation of the surface, separate coordinate systems are used for solid noise synthesis and rendering. A ray-surface intersection point in world space is first transformed into texture space before the above noise evaluation function is called. As will be explained in the next section, the local streamline integration and convolution are also performed in the texture space.

An important issue is how to decide or control the granularity or the frequency of the input noise. We can adjust the granularity of the noise simply through a scaling transformation in the texture space. Generally, the frequency of noise should be designed high enough relative to the resolution of vector field to provide a dense representation of a flow. On the other hand, an adequate frequency is always view dependent. The solid texturing technique includes a sampling process of the LIC texture on the 3D surface, and the Nyquist limit for sampling rate depends on the position of a surface in the world space. Since an LIC image is generated by applying a very thin one dimensional low pass filter along local streamlines to a high frequency white noise, the effect of blurring along the local streamline may be counteracted by the defect of aliasing, if the frequency of input white noise exceeds the Nyquist limit of the sampling rate. Given a set of viewing parameters, our algorithm calculates the Nyquist limit of sampling rate and use twice of its inverse as the default granularity of the white noise. The current implementation also allows the noise granularity to be adjusted interactively.

4   LIC in 3D Physical Space

As mentioned in Section 1, we assume a 3D surface is represented as a triangle mesh given as a list of triangles. Vectors together with other scalar values are stored at the vertices of the triangles. Our convolution algorithm is a 3D extension of the 2D LIC algorithm in Section 2. A ray-surface intersection point P in the world space is first transformed into a point $P(x, y, z)$ in texture space. Then the local streamline for the point is generated by integrating forward and backward along the direction of local vector in the texture space. As shown in Fig. 3, we treat the integer lattice for solid noise evaluation as a 3D extension of the 2D regular Cartesian vector field grid shown in Fig. 1.

In the original 2D algorithm, each step of the integration is performed in such a way that the streamline intersects one of the four edges of a square cell and enter into the adjacent cell(Fig. 1). Similarly here at each step of the integration, we let the streamline intersect one of the six facets of a hexahedral cell and enter into the adjacent cell in the integer lattice. To keep a streamline lying on the surface, the local vector is always projected onto the surface.

In other words, only the vectors tangential to the surface are used in the streamline calculation. Now the forward coordinate integration is given as follows:

$$P_0 = (x, y, z)$$

$$P_i = P_{i-1} + \frac{V_{P_{i-1}}}{|V_{P_{i-1}}|} \Delta s_{i-1}$$

$$P_0 = (x, y, z)$$

$$P_i = P_{i-1} + \frac{V_{\lfloor P_{i-1} \rfloor}}{|V_{\lfloor P_{i-1} \rfloor}|} \Delta s_{i-1}.$$

$$\Delta s_{i-1} = min(D(P_{i-1}, P_{rl}), D(P_{i-1}, P_{tb}), D(P_{i-1}, P_{fg}))$$

$$P_{rl} = \begin{cases} P_{right} & V^x_{\lfloor P_{i-1} \rfloor} > 0 \\ P_{left} & V^x_{\lfloor P_{i-1} \rfloor} < 0 \end{cases}$$

$$P_{tb} = \begin{cases} P_{top} & V^y_{\lfloor P_{i-1} \rfloor} > 0 \\ P_{bottom} & V^y_{\lfloor P_{i-1} \rfloor} < 0 \end{cases}$$

$$P_{fb} = \begin{cases} P_{front} & V^z_{\lfloor P_{i-1} \rfloor} > 0 \\ P_{back} & V^z_{\lfloor P_{i-1} \rfloor} < 0 \end{cases}$$

Here $V_{P_{i-1}}$ is the vector at the point $(P^x_{i-1}, P^y_{,i-1}, P^z_{i-1})$. Unlike the original 2D LIC algorithm, the vector at point P_{i-1} itself is used instead of the vector at the lattice point $(\lfloor P^x_{i-1} \rfloor, \lfloor P^y_{i-1} \rfloor, \lfloor P^z_{i-1} \rfloor)$. P_c(c: right,left,top,bottom,front,back) are the intersections between the streamline starting from the point P_{i-1} and the six facets of the cell constituted by the eight integer lattice points $(\lfloor P^x_{i-1} \rfloor + l, \lfloor P^y_{i-1} \rfloor + m, \lfloor P^z_{i-1} \rfloor + n), l, m, n = 0, 1$.

As shown in Fig. 3, the line passing through point P_{i-1} along the direction of the vector $V_{P_{i-1}}$ (the vector at P_{i-1}) may cross the border of the triangle enclosing P_{i-1} before it enters the adjacent cell of the integer lattice. In this case, we break off the integration at the intersection of the line and

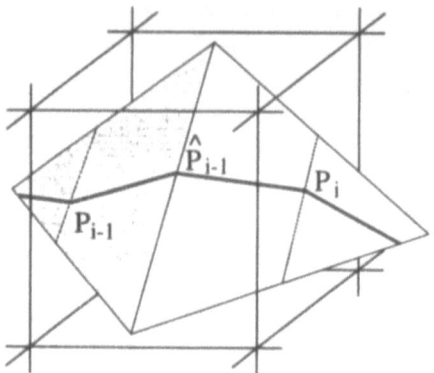

Fig. 3. Local streamline on a 3D surface

the edge of the triangle, and then restart the integration in the adjacent triangle with the vector at the intersection point $\hat{P}_{i-1}$. Such a process is repeated until the streamline intersects one of the six facets of the cell. Therefore, $D(P_{i-1}, P_c)(c : right, left, top, bottom, front, back)$ here are the sum of all line segments constituting the streamline from P_{i-1} to P_c and is usually larger than the Euclidean distances between P_{i-1} and P_c.

Now the convolution along the streamline at the intersection $P(x, y, z)$ can be represented as

$$V(x, y, z) = \frac{\sum_{i=0}^{l} Noise(\lfloor P_i \rfloor)h_i + \sum_{i=0}^{l'} Noise(\lfloor P_i \rfloor)h_i'}{\sum_{i=0}^{l} h_i + \sum_{i=0}^{l'} h_i'}$$

where

$$h_i = \int_{s_i}^{s_i + \Delta s_i} k(w) dw$$

Note that in case Δs_i is a streamline consisting of several line segments, h_i should be integrated piecewisely along each line segment. $Noise(\lfloor P_i \rfloor)$ is the solid noise value evaluated at integer lattice point $(\lfloor P_i^x \rfloor, \lfloor P_i^y \rfloor, \lfloor P_i^z \rfloor)$ through the function $Noise$ described in Section 3.

The vector at an arbitrary point on the surface is obtained by interpolating the vectors at the three vertices of the triangle enclosing the point. Area weighting interpolation [11] is used here. As shown in Fig. 4, the vector at a point P within a triangle $\triangle ABC$ is calculated in the following way:

$$V_p = \frac{V_A S_{BPC} + V_B S_{APC} + V_C S_{BPA}}{S_{ABC}}$$

Here, $S_{BPC}, S_{APC}, S_{BPA}$ and S_{ABC} are the area sizes of triangle $\triangle BPC$, $\triangle APC, \triangle BPA$ and $\triangle ABC$, respectively.

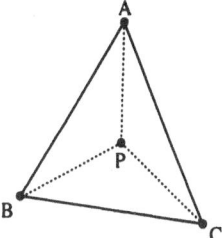

Fig. 4. Interpolation of vectors within a triangle

The length of convolution kernel l, l' is specified in unit cells of integer lattice. After the convolution has been performed, we transform the point back to world space and render it with the resulting LIC texture values.

5 Application

The new LIC algorithm has been implemented as a usr *module* of the commercial visualization software AVS [1]. The graphics workstation used was an SGI Indy with R5000 150MHZ CPU and 64 megabytes memory running an Iris 5.3 operating system. A set of widgets is provided for the interactive changing of various parameters such as the size of output image, viewing angles, the granularity of the noises and so on.

The first example is a visualization of the flow on a contour surface of velocity magnitude with a tornado flow dataset from Lawrence Livermore National Laboratory of the United States. The contour surface of velocity magnitude is generated using a modified Marching Cube program which also output the velocity at each vertices of triangles. The size of the original data set is $48 \times 48 \times 48$ and the contour surface is a mesh consisting 14950 triangles (Fig. 5). An LIC image generated with the new algorithm is shown in Fig. 6.

The second dataset is a numerical simulation of supersonic flow attacking a spiked blunt body [14]. The size of original curvilinear grid is $80 \times 30 \times 80$ and the second slice from the surface is used here (Fig. 7). Fig. 8-10 are the LIC images. In Fig. 8, the convolution integral has been normalized with a constant value independent of the length of local streamline. Thus the singularities, such as the region on the nose where the conical shock wave from the tip of the spike are reflected, are highlighted as a darker area than surroundings. Recall the algorithm given in Section 4, the streamline calculation only makes use of the tangential projection of vector on the surface and simply ignore the normal direction projections. In many cases, however, the flow in the normal direction of a surface is even more important than those in tangential directions. The image in Fig. 9(see Appendix) is generated by mapping the direction and magnitude of the vector in normal direction to the hue and saturation of color, respectively.

---

[1] AVS is a trademark of Advanced Visual Systems Inc.

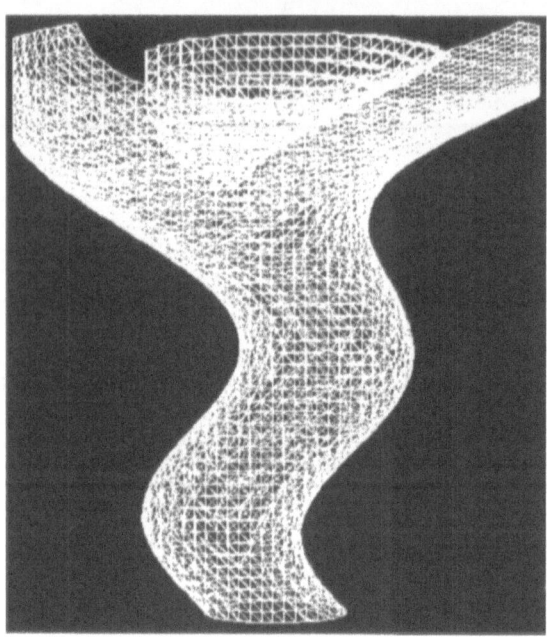

Fig. 5. The triangle mesh of velocity magnitude contour surface in the tornado dataset

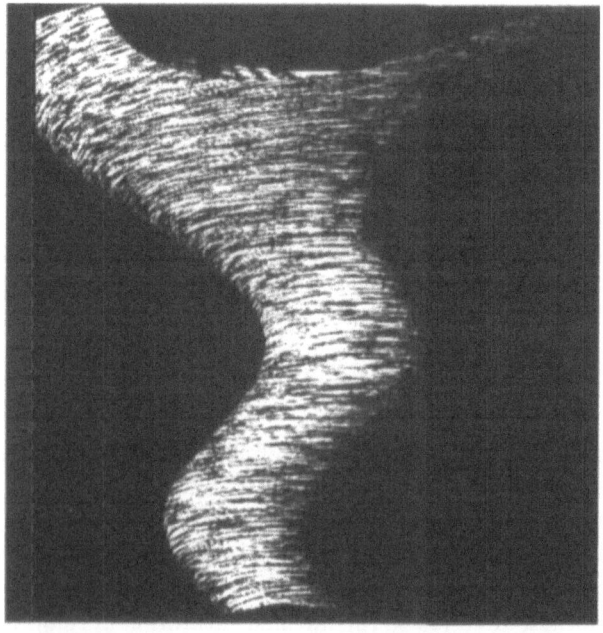

Fig. 6. The flow on the contour surface of Fig. 5

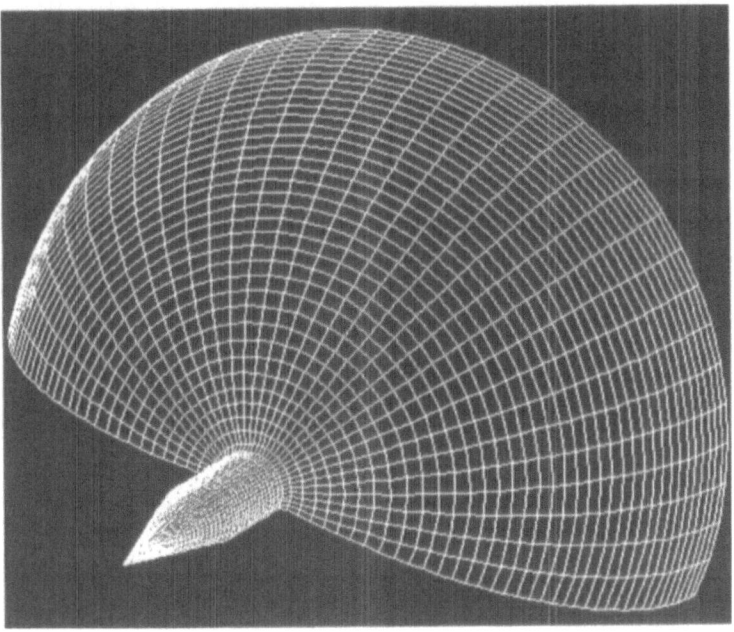

Fig. 7. The grid for the surface of the spiked blunt body

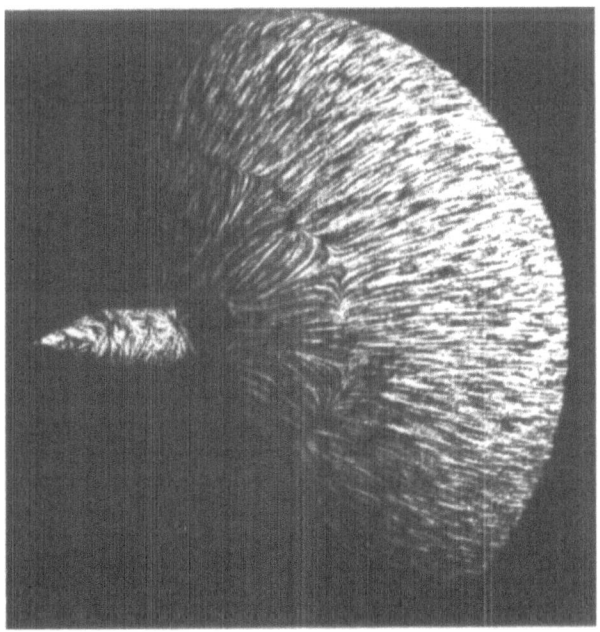

Fig. 8. The LIC image with constant normalization

The direction toward the surface is mapped to green and out of the surface is mapped to red. The low and high magnitude values are correspondently mapped to the low and high saturation values. Therefore a deep green color indicates the low flow toward the surface while a light red stands for a fast flow out of the surface. Fig. 10(see Appendix) is a close up view of the area near the base of the spike with the color showing the distribution of density.

As for the visualization of tangential vector magnitude, a mapping to local streamline length was not successful in case of the spiked blunt body. This is because the magnitude of vectors at some particular area, such as the head of the spaceplane and the tip of the spike, is extremely large compared with those at other area, and thus the directional blurring effect can hardly been seen in most area of the resulting LIC image. We found an animation generated by varying the amount of phase shift with the magnitude, as suggested by Forssell [2], is very effective in such case.

In the current implementation, we have focused mainly on exploring the effectiveness of the new algorithm in vector field visualization from the viewpoint of image quality. The algorithm has not been well tuned to achieve a good time efficiency. It took around 6 minutes to generate a 256 × 256 image for a mesh about of 10000 triangles. We found the ray-surface intersection test impact the execution time up to 60% especially in case of a large triangle mesh. We are now experimenting with a fast ray-surface intersection algorithm by utilizing hardware Z-buffer and we believe the rendering speed should be drastically improved with the new implementation.

6   Concluding Remarks

A new LIC algorithm for visualizing the flow on 3D surfaces has been presented. With the use of solid texturing, the algorithm is applicable to any complex 3D surfaces to generate LIC images without distortion. One of the most attractive features of a texture-based approach is its generality and simplicity. Our solid texture-based algorithm inherits such a property of 2D algorithms and can be easily realized in any existing environment, such as a personal computer without the supporting of texture mapping hardware.

Needless to say, realtime rendering which enables users to view their data interactively from different angles and positions is very important in scientific visualization. One of our major research directions will be concerned on this aspect. Antialiasing is another important and challenging issue. The aliasing defect can be badly amplified when performing phase-shift animation. Combining the proposed technique with other existing visualization tools may open up new applications for the technique. For example, showing the local flow features on a streamsurface or stream ribbon with our LIC algorithm can produce useful images similar to that generated by the surface particle technique [9].

7 Acknowledgements

The authors are deeply grateful to Kozo Fujii from Japan Institute of Space and Astronautical Science, Isamu Kuroki and Hideo Miyachi from Kubota Graphics Technology Inc., Henwei Shen from MRJ Inc. at NASA Ames Research Center, and Nelson Max from Lawrence Livermore National Laboratory for their cooperations in providing the datasets. Special thanks to Yasunori Funato, member of our laboratory for his help in implementation and preparing of manuscripts. Finally we would like to thank Issei Fujishiro from Ochanomizu University for his helpful comments and continual support to the work.

References

1. Cabral, B. and Leedom, C.: Imaging Vector Field Using Line Integral Convolution. Proceedings of SIGGRAPH'93, ACM SIGGRAPH, pp. 263-270 (1993).
2. Forssell, L. K. and Cohen, S. D.: Using Line Integral Convolution for Flow Visualization: Curvilinear grids, Variable-speed Animation and Unsteady Flows. IEEE Transaction on Visualization and Computer Graphics, Vol.1, No.2, pp. 133-141 (1995).
3. Leeum, W. C. and Wijk, J. J. van: Enhanced Spot Noise for Vector Field Visualization. Proceedings of Visualization'95, pp. 233-239 (1996).
4. Lewis, J. P.: Algorithms for Solid Noise Synthesis. Computer Graphics, Vol. 23, No.3, pp. 263-270 (1989).
5. Lorensen, W. E. and Cline, H. E.: Marching Cubes: A high Resolution 3D surface Construction Algorithm. Computer Graphics, Vol. 21, No.4, pp. 163-169 (1987).
6. Max, N., Crawfis, R. and Grant, C.: Visualizing 3D Velocity Fields near Contour Surfaces 3D Scalar Functions. Proceedings of Visualization'94, pp. 248-255 (1994).
7. Peachey, D. R.: Solid Texturing of Complex Surfaces. Computer Graphics, Vol. 19, No.3, pp. 279-286 (1985).
8. Perlin, K.: An Image Synthesizer. Computer Graphics, Vol. 19, No.3, pp. 287-296 (1985).
9. Post, F.H. and Walsum, T. van: Fluid Flow Visualization. Visulization'93 Tutorial 3, pp. 1-37 (1993).
10. Shen, H., Johnson, C. R. and Ma, K.: Visualization Vector Fields Using Line Integral Convolution and Dye Advection. Proceedings of the 1994 Symposium on Volume Visualization, pp. 63-70 (1996).
11. Sadarjoen, A., Walsum, T. van, Hin, A. J. S. and post, F. H.: Particle Tracing Algorithms for 3D Curvilinear Grids. 5th Eurographics Workshop on Visualization in Scientific Computing, (1994).
12. Stalling, D. and Hege, H. C.: Fast and Resolution Independent Line Integral Convolution. Proceedings of SIGGRAPH'95, ACM SIGGRAPH, pp. 249-256 (1995).
13. Wijk, J.J. van: Spot Noise: Texture Synthesis for Data Visualization. Computer Graphics, Vol. 24, No.4, pp. 309-318 (1991).
14. Yamauchi, M., Fujii, K. and Higashino, F.: Numerical Investigation of Supersonic Flows around a Spiked Blunt-Body. Journal of Spacecraft and Rockets, Vol. 32, No.1, pp.32-42 (1994).

Editors' Note: see Appendix, p. 178 for colored figures of this paper

A Framework for Physically-Based Information Visualization

T. C. Sprenger, M. H. Gross, A. Eggenberger, M. Kaufmann[*]
Computer Science Department
Swiss Federal Institute of Technology
ETH Zürich, Switzerland

[*]Swiss Bank Corporation, IT-Camp Basel, Switzerland

Abstract The following paper describes a framework for the visualization and analysis of economic data. It can be employed in the context of risk analysis, stock prediction and other tasks being important in the context of banking. The system bases on a quantification of the similarity of related objects, which governs the parameters of a mass-spring system, organized as two concentric spheres. More specifically, we initialize all information units onto the surface of the inner sphere and attach them with springs to the outer sphere. Since the spring stiffnesses correspond to the computed similarity measures, the system converges into an energy minimum, which reveals multidimensional relations and adjacencies in terms of spatial neighborhoods. In order to simplify complex setups we propose an additional clustering algorithm for postprocessing. Furthermore, depending on the application scenario we support different topologic arrangements of related objects. In addition, we implemented various interaction techniques allowing semantic analysis of the underlying data sets. The versatility of our approach is illustrated by two examples, namely a comparison of agricultural productivity and an analysis of the relation between interest rates and other economic data.

1 Introduction

The visualization of complex, multidimensional, non-numeric information and of their relationships is an emerging subfield of increasing importance in scientific visualization. Nowadays, global computer networks and distributed data bases, such as the world wide web (WWW), provide platforms for new dimensions of retrieval systems for information units. As a consequence, the scientific visualization and computer graphics communities have been challenged to develop advanced tools for understanding, navigating and interactively analyzing the associated information spaces. However, as opposed to most of the classical data sets in scientific visualization, information spaces carry over entirely new qualities of problems. The most important ones can be summarized as follows:

- *Multidimensional relationships:* information units are generally related to many other units. The resulting topological organization corresponds to a multidimensional graph. Thus, adjacencies cannot be visualized straightforwardly and have to be mapped into subspaces. Here, we can carry over some interesting methods being already used in graph layout.

- *Measuring similarity:* In contrast to many scientific data sets, information space is an abstract entity and there is no specific reason to employ Euclidean metrics to project similarities into a 3 dimensional subspace. Moreover, up to now there is no mathematical framework or paradigm, on how to map scores and similarities provided by retrieval systems onto a model in a three-dimensional world.

- *Clustering and hierarchies:* The huge amount of information forces the use of a multiresolution setup. Hence, appropriate methods for the clustering of objects and for interactive level of detail control are needed.

Due to the importance of information visualization for many applications, various interesting approaches can be found in literature and excellent surveys are available [20]. [5], [14] for instance, visualized text documents and clusters as galaxies and themescapes, whereas [4] proposed cone trees which specifically address hierarchical organization. Another promising method is [12] or [16], who essentially used self-organizing schemes and neural networks to arrange information objects of the WWW. In a more general understanding, multidimensional visualization problems have been stressed in [6] or [1]. Here, mathematical projection algorithms were introduced to map data into subspaces, while preserving their most important features. Interestingly, many current methods use physically based paradigms, such as [18] or [13], where information units are taken as nodes of some generalized mass spring system revealing the structure of relations upon relaxation. These types of multidimensional visualization methods have been studied extensively in graph theory, and efficient algorithms had been introduced for fast graph relaxation, such as [11] and [12].

The work reported in our paper was inspired by the research summarized above. However, unlike existing methods, our approach was mostly application driven, where the context was focussed on visualization problems arising in financial service providing. Therefore, we define a propriety mathematical framework for quantifying relationships in information space. In addition, we propose a visualization paradigm that considers all information units as initially located on the inner part of two concentric spheres and as attached with springs to the outer one. The strengths of relations of different objects are correlated to the stiffnesses of springs between them. After initialization, the system converges into an equilibrium stage by solving the underlying differential equations using popular strategies [2], [15]. Thus, the energy minimum represents spatial adjacencies of objects which are similar to each other in information space. Moreover, since the similarity matrix also encodes the topology of the underlying graph, standard algorithms from graph theory can be employed to discover the indirect links of objects and to find minimal paths between them. In order to simplify the geometric complexity of large scale data sets, we propose an additional clustering by computing an ellipsoidal hull around individual objects. The ellipsoid is parametrized by the principal components of the underlying cluster.

The organization of the paper is as follows: First we introduce the mathematical definition of similarity and explain our metric. Section 3 discusses the paradigm of two concentric spheres and describes the strategies for initial positioning and clustering of information units. In section 4 interactive analysis algorithms are elucidated. Finally, we illustrate the performance of our method by two examples: a comparison of agricultural productivity and an analysis of the relation between interest rates and other economic parameters.

2 Mathematical Foundations

This section introduces the mathematical foundations required to understand the approach. First, our metric for similarity in information space is elaborated, since it represents a major prerequisite for visualization. Furthermore, we briefly review the principles of the dynamics of mass spring and particle systems.

2.1 Measuring Similarity in Information Space

One of the very challenging problems of information visualization is the definition of a mathematical framework for the quantification of similarity of entities in information space. We decided to found our framework on vector spaces, but it also supports probabilistic, Boolean and Euclidean approaches for information retrieval. Specifically, we assume the metric as being computed in a preprocessing step, and providing the input parameters of a physically-based system.

Recalling some foundations of data base research we formalize the information retrieval process as follows: Let $A=\{A_1, .., A_k\}$ be a set of attributes (key words) and let's consider n objects $O=\{O_1, .., O_n\}$, where each object O_i is assigned to a score vector s_i of dimension k. The component s_{il} represents the relative importance of document O_i with respect to key word A_l and is assumed to be bounded by $[0, 1]$.

We now define the so-called *similarity* c_{ij} of two objects O_i and O_j by the dot product of the associated scores normalized with respect to the dimension k:

$$c_{ij} = c_{ji} = \sqrt{\frac{s_i \bullet s_j}{k}} = \sqrt{\frac{1}{k}\sum_{l=1}^{k} s_{il}s_{jl}} \quad i, j = 1...n \quad (1)$$

The resulting elements $0 \leq c_{ij} \leq 1$ form a symmetric, positive definite $n \times n$ *similarity matrix* C:

$$C = \begin{bmatrix} c_{11} & c_{12} & \cdots & c_{1n} \\ c_{21} & c_{22} & \cdots & c_{2n} \\ \cdots & \cdots & \cdots & \cdots \\ c_{n1} & \cdots & \cdots & c_{nn} \end{bmatrix} \quad (2)$$

where $c_{ij} = c_{ji}, 1 \leq i, j \leq n$.

The *self-similarity* c_{ii} of an object O_i is provided by the normalized length of the score vector

$$c_{ii} = \frac{|s_i|}{\sqrt{k}}, i = 1...n \quad (3)$$

Note, that C also reflects whether or not two objects are related. Hence, it can be considered as some sort of generalization of the so-called adjacency matrix, well-known from graph theory [8]. The similarity defined by (1) can be interpreted as the projection of one score vector onto another. Needless to say that the quality of the selected key words and retrieval algorithms is crucial for the quality of our similarity matrix.

The matrix from above quantifies the strengths of relations of information units in an abstract space. At this point in time, we have to analyze the associated topology by which adjacent objects and their relations are described. We end up in a *network type model* or graph, such as illustrated in Figure 1a, which is obviously one of the fundamental topologies for visualizing related information units.

However, if we target at visualizing how particular objects match a predefined string of key words, the upper model is apparently not well shaped. Moreover, a new *optimal* object O_{n+1} has to be added, whose scores are all set to: $s_{(n+1),l} = 1, l = 1,..,k$. The similarity matrix is constructed by computing the dot products between all objects and the one newly inserted O_{n+1}. That is, we relate all objects to a theoretically optimal score vector. The structure of the associated $(n+1)x(n+1)$ matrix reveals only elements on the diagonal and in the last column and row, respectively.

$$\mathbf{C}_{star} = \begin{bmatrix} c_{11} & 0 & ... & c_{1(n+1)} \\ 0 & ... & ... & ... \\ ... & 0 & ... & c_{k(n+1)} \\ c_{(n+1)1} & ... & ... & c_{(n+1)(n+1)} \end{bmatrix} \tag{4}$$

$c_{ij} = 0, (i \neq j) \wedge 1 \leq i, j \leq n$ and $0 \leq c_{i(n+1)} = c_{(n+1)i} \leq 1, 1 \leq i \leq n$.

Based on the metric defined by (1) the resulting elements compute to

$$c_{(n+1)i} = c_{i(n+1)} = \sqrt{\frac{1}{k} \sum_{l=1}^{k} s_{il}} \tag{5}$$

As a consequence, the underlying topology has changed and converts into a *star type* arrangement, such as depicted in Figure 1b. The central object is supposed to be the theoretical optimum and taken as a fix point. It is clear that a combination of both *network type* and *star type* topology is accomplished straightforwardly. Formally, we can also construct a third arrangement like the one of Figure 1c. It enables us to visualize both relations of objects between each other and with respect to a predefined score. A detailed elaboration of the associated visualization paradigm will be given in the next section.

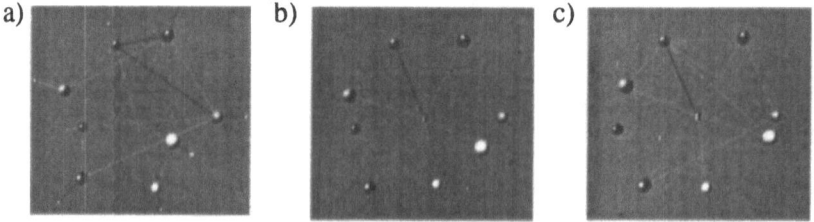

a)　　　　　　　b)　　　　　　　c)

Fig 1. Different type of visualization models for multidimensional relations in information space: a) Network b) Star c) Combined Setting

2.2 Mass-Spring Systems

Mass-spring systems are linear finite elements and have been used widely and successfully in computer graphics [2], since the underlying physics is based on linear differential equations and is straightforward to implement. The principal equation that governs the attractive force $\mathbf{F}_{ij}$ between two attached masses m_i and m_j at spatial locations $\mathbf{r}_i$

and $\mathbf{r}_j$ and velocities $\mathbf{v}_i$ and $\mathbf{v}_j$ is given by

$$\mathbf{F}_{ij} = c_{ij}(\mathbf{r}_j - \mathbf{r}_i)\left(1 - \frac{l_0}{|\mathbf{r}_j - \mathbf{r}_i|}\right) + f(\mathbf{v}_j - \mathbf{v}_i) \tag{6}$$

where c_{ij} stands for the spring stiffness, l_0 represents a bias length and f is the friction.

Applying Newton's law converts (6) into a second order linear differential equation. Note, however, that a straightforward discretization and integration results in algorithms of quadratic complexity. Therefore, it is recommended to invoke more sophisticated techniques, such as the ones proposed in [14] or [12].

3 Arranging Information on a Sphere

In the following section we describe the visualization paradigm which combines the metric introduced earlier with the physically based mass-spring approach. More precisely, we assign the computed similarities to the individual stiffnesses of springs linking information units to each other. Hence, we end up with an intermediate mapping of similarities onto parameters of a physically-based system. This approach has already proved to promise good results [18], [13] and [14]. In this context we address two major novelties of our framework: topology and arrangement of objects at initialization.

3.1 The Paradigms

Given a set of documents $O_1,..,O_n$ the visualization method employs two concentric spheres for the initial positioning of information units as illustrated in Figure 2a. The objects are placed on the surface of the inner sphere and are attached with springs to the *virtual* outer sphere and to each other. The major advantage of this arrangement is the degree of symmetry inherent to the geometry of a sphere of radius R. Consequently, the model handles our three fundamental topologies from Figure 1.

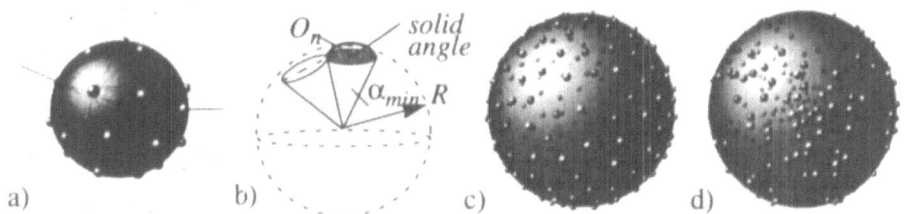

Fig 2. a) Initialization of information objects on a virtual sphere b) Poisson disc sampling for initial positioning c) Poisson disc distribution for: $f_d = 1.0$, d) $f_d = 0.3$

3.2 Initialization Procedures

Positioning of Objects

First, for a given set of n objects, the initial positions on the spherical surface have to be found. A straightforward positioning could be at random. However, due to the problems arising with small numbers of objects, we recommend a Poisson disc sampling procedure of the surface [9]. As depicted in Figure 2b each object is assigned to a solid angle element and the distance of two objects must not exceed a distance threshold d_{min}, computed by the following relation:

$$d_{min} = 2Rf_d\sin\left(\frac{\alpha_{min}}{2}\right) = 4R4\sqrt{1 - \frac{f_d}{n}}\sqrt{\left(1 - \sqrt{1 - \frac{f_d}{n}}\right)} \tag{7}$$

The influence of the factor f_d on the uniformity of the distribution of the initial positions is illustrated in Figure 2c and Figure 2d.

We have to be aware that our model corresponds to a multidimensional graph visualization problem [11] and is prone to all problems associated with subspace procedures [1]. Consequently, in order to prevent the system from convergence to a local minimum, tightly related objects should be positioned as close as possible to each other upon initialization. It is clear, that we won't succeed in the general case, but the traversal strategy reported below avoids most initial ill-conditioning.

We start with an initial weighting of all objects, where the weight w_i is defined by the sum of all adjacent similarities.

$$w_i = \sum_{j=1}^{u_i} c_{ij} \quad ; \quad u_i: \text{number of objects adjacent to } O_i \tag{8}$$

This weighting emphasizes the importance of units with strong relations. From there, a list $O=\{O_1,..,O_n\}$ of all objects is built which is sorted with respect to the weights w_i. The assignment of information objects to the computed surface positions is figured out by a *breadth-first* strategy. The algorithm starts from the most important object and assigns first positions to all directly linked objects. These positions are ranked according to their distance from the initial one in 3D space. From there, the procedure traverses the list recursively until all objects are assigned. A pseudocode fragment for the method is given by:

```
O={O₁,...,Oₙ}        // initial object list //
P={P₁,...,Pₙ}        // initial positions //
T={}                 // list of sublists //
while (O not empty) do
    fetch object Oᵢ from O | wᵢ = wₘₐₓ
    assign random position Pₖ to Oᵢ: Oᵢ ->Pₖ
    remove Pₖ from P
    generate list Lᵢ of all directly
    linked objects O₁ ∈ O sorted with respect to cᵢ₁
    keep wᵢ with Lᵢ
    add Lᵢ into T
    while (T not empty) do
        fetch sublist Lₘ from T | wₘ = wₘₐₓ
        remove all objects ∈ Lₘ from O
        while (Lₘ not empty) do
            fetch next object Oⱼ from Lₘ
            assign free position Pₖ to Oⱼ |
            dist (Pₖ,Pₘ) = min: Oⱼ -> Pₖ
            remove Pₖ from P
            generate list Lⱼ of all directly
            linked objects O₁ ∈ O sorted with
            respect to c₁ⱼ
            keep wⱼ with Lⱼ
            add Lⱼ into T
```

od
od
od

It should be stated, that a *depth-first* traversal had been implemented as well but experimental results perform similarly.

Spring Stiffness Assignment

Another important aspect is the computation of the stiffnesses of the anchor springs attaching the objects to the outer sphere. It is clear that we have to take into account the individual connectivity of a single information unit. In order to understand the computation, one should be aware that the anchor spring influences the degree to which closely related objects converge to each other during energy minimization. Therefore, we propose the following method, where, for convenience, all relations hold for $t = 0$:

Let l_0 be the initial length of the anchor spring i and let F_a^i be its attractive force. The resulting force $F_h^i = \sum_{j=1}^{u_i} F_{ij}$ is computed from all springs linking object $O_j = m_j$ to others according to the adjacency matrix. This is shown in Figure 3a.

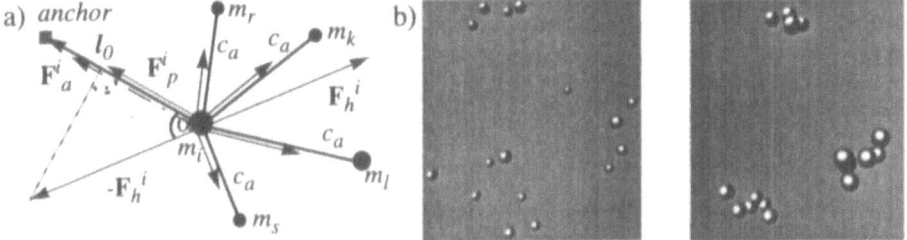

Fig 3. a) Spatial arrangement of the resulting force vector for initial spring stiffness computation
b) Clustering of information units for different cluster-factors: $f_c = 0.9$ c) $f_c = 0.2$

The computation of the anchor spring stiffness is figured out by projecting $-F_h^i$ onto the vector l_0, which defines the direction of the anchor spring. Introducing an additional factor f_c, we set

$$|F_a^i| = |l_0|c_j = f_c|F_p^i| = f_c\cos\alpha|F^i|$$ (9)

and obtain the relation for the required anchor spring stiffness c_i:

$$c_i = \frac{f_c l_0 \bullet (-F_h^i)}{l_0^2}$$ (10)

In order to compute F_i^h all springs are supposed to have a uniform stiffness given by averaging:

$$c_a = \frac{1}{u_i}\sum_{j=1}^{u_i} c_j$$ (11)

Experiments have shown that this approach is superior to the immediate usage of the individual spring stiffnesses, since it balances differences of individual spring lengths. Note, that the same relations hold for the star type topology.

The factor f_c can be considered as a clustering factor, which takes immediate influence on the degree to which related objects converge to each other during simulation. Figure 3b and Figure 3c depict results of an experimental setup for different values of f_c.

4 Exploration of Information Space

Further important issues to be addressed concern the graph topology of related documents in information space. This formalism enables us to carry over some of the fundamental algorithms [8] to support interactive analysis and clustering, where in addition a PCA based clustering mechanism allows to simplify the structure of complex subregions. Minimal path procedures allow discovery and quantification of indirect links between objects.

4.1 Discovering Relations using Minimal Paths

For some applications it is useful to explore indirect relationships rather than immediate ones. This holds also for probabilistic settings where the links and adjacencies are marked by probabilities $p<1$. In order to provide tools for interactive exploration we take advantage of the graph structure of the visualization problem. Here, the indirect links which optimize a specific cost function are interesting for analysis.

Let $c_{ij} = p_{ij}$ be the probability that ranks the strength of two related objects O_i and O_j. Assuming statistical independence, we compute the probability p_{acc} of two objects O_S and O_E indirectly linked through a specific path by:

$$p_{acc} = \prod_{i, j \in path(S, E)} p_{ij} \tag{12}$$

Formally, paths from one node to another are derived from the *transitive closure* in graph theory, a Boolean matrix, whose elements are either 1, if two nodes are connected or 0 if not. Since these types of algorithms belong to the standard repertoire of computer science textbooks, we refer to those [10].

4.2 Clustering and Level-of-detail (LOD)

In order to simplify the geometry and topology of complex object arrangements it is necessary to provide an efficient level-of-detail strategy. Initial work for information visualization is reported in [7] who accomplished simple clustering by wrapping hyperspheres around groups of objects. The transparency of the hyperspheres was controlled as a function of the distance to the viewer. Unlike this approach we propose a K-means and PCA based clustering mechanism [19] which will be explained in the upcoming section.

The basic idea is to wrap ellipsoids around each cluster whose shape is controlled by the principal components of the respective cluster. The method is designed as a two pass procedure, where in a first step all objects in the scene are divided into a set **K** of disjoined subsets. The algorithm passes though all objects and assigns a new cluster if the distance of the current object to all existing clusters exceeds a threshold *delta*. This variable finally governs the granularity of the generated clusters. A pseudocode fragment is given below:

```
O={O₁,..,Oₙ}  // initial object list; rᵢ = position of Oᵢ //
K={}          // set of clusters; mⱼ = centroid of cluster Kⱼ //
while (O not empty) do
    fetch object Oᵢ from O | minima = delta
        iterate over all clusters Kⱼ ∈ K do
            if (|mⱼ - rᵢ| < minima) then
                Kₘᵢₙ = Kⱼ
                minima = |mⱼ - rᵢ|
            fi
        od

        if (minima < delta) then
            add Oᵢ into Kₘᵢₙ | update mₘᵢₙ
        else
            create new cluster Kₙₑw
            add Oᵢ into Kₙₑw | mₙₑw = rᵢ
            add Kₙₑw into K
        fi
    od
end
```

The second pass comprises the parametrization of an affine map which transforms the initial 3D ellipsoidal shape appropriately into the scene. For cluster K_j this transform is defined by a translation vector $\mathbf{m}_j$, a scaling matrix $\mathbf{S}_j$ and a rotation matrix $\mathbf{R}_j$. Let n_j be the number of objects in cluster K_j we obtain the required translation vector immediately as the centroid of the cluster:

$$\mathbf{m}_j = \frac{1}{n_j} \cdot \sum_{i=1}^{n_j} \mathbf{r}_i \quad ; \quad \mathbf{r}_i: \text{object positions in cluster } K_j \tag{13}$$

In addition, the 3x3 Covariance-matrix of the cluster is given by

$$\mathbf{M}_j = \frac{1}{n_j} \cdot \sum_{i=1}^{n_j} (\mathbf{r}_i - \mathbf{m}_j) \cdot (\mathbf{r}_i - \mathbf{m}_j)^T \tag{14}$$

By solving the Eigenproblem $\mathbf{M}_j \cdot \mathbf{u}_k = \sigma_k \cdot \mathbf{u}_k$ we compute the 3 Eigenvalues σ_{j1}, σ_{j2}, σ_{j3} and the associated Eigenvectors $\mathbf{u}_{j1}$, $\mathbf{u}_{j2}$, $\mathbf{u}_{j3}$ which define the required transformation matrices, where

$$\mathbf{S}_j = \begin{bmatrix} \sqrt{\sigma_{j1}} & 0 & 0 \\ 0 & \sqrt{\sigma_{j2}} & 0 \\ 0 & 0 & \sqrt{\sigma_{j3}} \end{bmatrix}, \quad \mathbf{R}_j = \begin{bmatrix} u_{j1x} & u_{j2x} & u_{j3x} \\ u_{j1y} & u_{j2y} & u_{j3y} \\ u_{j1z} & u_{j2z} & u_{j3z} \end{bmatrix} \tag{15}$$

Note that the 3x3 Eigenproblem can be solved analytically.

Thus the transform is figured out by the following set of equations. We start from the implicit equation of the unit sphere with surface vector $\mathbf{x}_j$:

$$\mathbf{x}_j = (x, y, z) \mid x^2 + y^2 + z^2 = 1 \tag{16}$$

and perform a subsequent affine mapping by

$$\mathbf{x}_j^e = \mathbf{m}_j + \mathbf{R}_j \cdot \mathbf{S}_j \cdot \mathbf{x}_j^0 \quad ; \quad \mathbf{x}_j^e: \text{surface vector of the ellipsoid} \tag{17}$$

Due to the statistical properties of the principal components it is not guaranteed that all objects of a cluster are enclosed by the ellipse. Thus we carry out additional postprocessing and grow the hull until all objects are enclosed.

5 Examples

The following section illustrates the performance and versatility of our approach by using two different examples: the visualization of agricultural productivity of selected countries and the visualization and analysis of interest rates correlating with other important economic parameters. Note, however, that the 2D pictures of this section do not reveal the full 3D arrangement computed by our method [21].

5.1 Agricultural Productivity

A classical example for visualization and analysis of multidimensional relations is given by taking some items from the world market. We employed figures from the yearly production of different agricultural products for selected countries, some of which are listed in Table 1.

Table 1: Productions of different countries in 1994 (subset)

Product Country	milk	meat	wheat	rice	potatoes
USA	69250	32091	63133	8547	19050
Canada	7750	2989	23180	0	0
China	5610	41424	102005	175608	36160
India	32112	4117	57802	117600	16318
Brazil	15774	8082	2127	11166	0
France	24900	6179	29944	0	4903
Germany	28200	5772	16100	0	12260
Switzerland	3300	448	0	0	800
....	....	....	....	....	....
Total	460058	193809	534301	531341	283306

Set of products **A** = {*butter, milk, meet, wheat, rice, soja, potatoes, sugar, bananas, cacao, coffee, tea*}.

Set of countries **O** = {*USA, Canada, China, India, Pakistan, Vietnam, Sri Lanca, Russia, Romania, Brazil, Argentina, Cuba, Columbia, Costa Rica, Peru, Ecuador, France, Netherlands, Italy, Austria, Germany, Switzerland, Australia, Japan*}.

A can be considered as the set of key-words and **O** as the set of information units. The figures of Table 1 were taken as a basis to construct a similarity matrix **C** according to the metric of section 2.

Figure D and Figure E (see Appendix) show the results of the *network type* model approach by contrasting the initial setup to the energy minimum of the particle model. All objects are textured according to their regional location, the USA and Switzerland are presented by their flags. Line color and thickness reflect the strength of the connection. We observe that most objects are arranged around the USA which forms a cluster center due to its high productivity. Conversely, Switzerland, as a small country moved apart. Furthermore, the results of the clustering algorithm are presented, where the clusters are visualized as transparent hulls whose opacity is controlled by the distance to the camera position.

Further interactive analysis is depicted in Figure F (see Appendix) where all objects with a direct link to the USA are presented. Those countries also lay within a particular radius. Changing this threshold reveals gradually those countries competing immediately to the USA on the world market. The strengths of the competition is approximated by the elements c_{ij} of **C**, also indicated in the figure.

5.2 Long Term Interest Rates

In the second example we contrast our method with a traditional way of analyzing multidimensional relationships of economic indicators. The goal is to evaluate the influence of the indicators presented in the diagrams of Figure 4a on the long term interest rates of individual countries. Each of these indicators was computed relative to the USA as a reference. The state of the art approach, as depicted in Figure 4b, consists of producing bar charts showing the correlation with individual indicators for different countries. These charts form a basis for further interpretation performed by the financial analyst. In order to map the problem onto our visualization paradigm we start from a special instance of the *network type* model. By imposing displacement constraints we first generate a subset of objects which keep their position during relaxation. For visualization, we map our indicators on these object types. Conversely, we drop the anchor stiffnesses of all other objects to zero, that is we cut off their anchors. These freely movable objects represent the countries and are connected via links to all rigid objects from above. The spring stiffness of a link conforms to the correlation of the associated indicator to the long term interest rates of this country. Note that the movable objects are not interconnected.

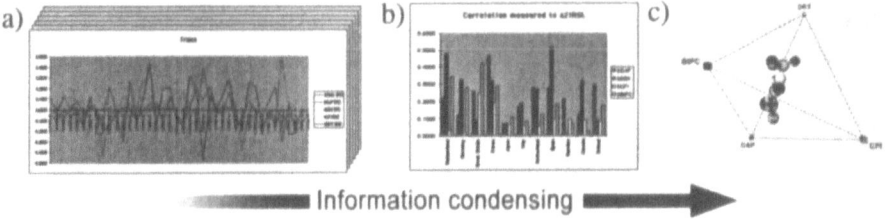

Fig 4. Condensing multivariante relationships: a) Stack of conventional diagrams b) Correlation tables c) Physically-based visualization paradigm

Figure G and Figure H (see Appendix) display two views on the relaxed model. The cubes at the vertices of the tetrahedral structure stand for the different indicators taken into account and the spheres representing the countries are textured with their flags. Although the definition of individual indicators is beyond the scope of this paper we observe that the interest rates of Canada correlate tightly with the index DRX, whereas Switzerland relates more closely to GAP and CPI. Conversely, Germany is located near the center of gravity of the plane spanned by DRX, BIPC and GAP and is hence equally influenced by those.

Conclusions and Future Work

We presented a new variant for physically-based information visualization and illustrated its versatility. The fundamental idea is to arrange all information units on the inner part of two concentric spheres and to attach them with springs to each other. Relaxation of the model figures out the structural relations in information space. Specifically, we are convinced that the physically-based approach fits nicely advanced I/O concepts with force and tactile feedback. Future research has to encompass a generalization of the physically-based approach including timeseries, limited lifetime of particles and advanced clustering algorithms using isosurfaces.

References

1. M. Gross, R. Koch: Visualization of Multidimensional Shape and Texture Features in Laser Range Data using Complex-Valued Gabor Wavelets, IEEE-Transactions on Visualization and Computer Graphics, Vol. 1, No. 1, pp.44-59, 1995

2. A. Witkin, D. Baraff, M. Kass: Physically-Based Modeling, SIGGRAPH tutorial course Notes No. 34, 1995

3. M. Harada, A. Witkin, D. Baraff: Interactive Physically-Based Manipulation of Discrete/Continuous Models, Proceedings Siggraph 95, pp. 199-208, 1995

4. J. Carriere, R. Katzman: Research Report - Interacting with Huge Hierarchies: Beyond Cone Trees, Proceedings of the IEEE Info. Vis. 95, pp. 74-81, 1995

5. J. Wise, et al.: Visualizing the Non-Visual: Spatial analysis and Interaction with Information from text Documents, Proceedings of the IEEE Info. Vis. 95, pp. 51-58, 1995

6. F. Young, P. Rheingans: Visualizing Strucutre in High-Dimensional Multivariate Data, IBM Journal of Research and Development, Vol. 35, No. 1/2, pp. 97-107, 1991

7. R. Hendley, et al.: Case Study - Narcissus: Visualizing Information, Proceedings of the IEEE Information Visualisation 95, pp. 90-96, 1995

8. J. Nievergelt, K. Hinrichs: Algorithms and Data Structures with Applications to Graphics and geometry. Englewood Cliffs: Prentice Hall, 1993

9. A. S. Glassner: Principles of Digital Image Synthesis, Morgan Kaufmann Publishers,San Francisco, 1995

10. T. H. Cormen, C. E. Leiserson, ans R. L. Rivest: Introduction to Algorithms, MIT Press, Cambridge, Massachusetts, 1994R. Koch, M. Gross, et al.: Simulating Facial Surgery Using Finite Element Models, Proceedings of SIGGRAPH 96, pp. 421-428, 1996

11. A. Frick, A. Ludwig and H. Mehldau: A fast adaptive layout algorithm for undirected graphs, Proceedings of Graph Drawing 94, LNCS 894, Springer Verlag 1995

12. I. Bruss, A. Frick: Fast Interactive 3-D Graph Visualization, Proceedings of Graph Drawing 95, Springer Verlag, LNCS 1027, p. 99-110

13. R. J. Hendley, N. S. Drew: Visualisation of complex systems, http://ww.cs.bham.ac.uk

14. M. Chalmers: A Linear Time Layout Algorithm for Visualizing High-Dimensional Data, Proceedings of the IEEE Information Visualization 96, pp. 127-132, 1996

15. A. Witkin: Particle System Dynamics, SIGGRAPH 96 Course Notes 34, pp C1-C12, 1996

16. A. Wood, et al.: HyperSpace: Web Browsing with Visualisation. Third International World-Wide Web Conference, Poster Proceedings, Darmstadt, Germany, pp. 21-25, 1995

17. T. R. Henry, S. E. Hudson: Interactive Graph Layout, Proceedings of the ACM SIGGRAPH Symposium, Proceedings ACM Siggraph Symposium on UI Software, 1991

18. C. L. Bentley: Animating Multidimensional Scaling to Visualize N-Dimensional Data Sets, Proceedings of the IEEE Information Visualisation 96, pp. 72-73, 1996

19. K. Fukunaga: Introduction to Statistical Pattern Recognition. 2nd Edition, New York: Academic Press, 1990

20. S. Card, S. G. Eick, N. Gershon: Information Visualization, SIGGRAPH 96 Course Notes 8, 1996

21. CGRG Homepage, ETH Zürich, http://www.inf.ethz.ch/department/IS/cg/html/research/infovis.htm

22. M. H. Gross, T. C. Sprenger, J. Finger: Visualizing Information on a Sphere, Technical Report, ETH Zürich, 1997

Editors' Note: see Appendix, p. 179 for colored figures of this paper

Feature Extraction from Pioneer Venus OCPP Data

Freek Reinders [1]   Frits H. Post [1]   Hans J.W. Spoelder [2]

[1] Delft University of Technology
Faculty of Technical Mathematics and Informatics
PO Box 356, 2600 AJ Delft, The Netherlands

[2] Free University Amsterdam
Faculty of Physics and Astronomy
De Boelelaan 1081, 1081 HV Amsterdam, The Netherlands

Abstract. Scientific visualization provides means to explore data and highlight interesting features in the data. In this paper we will discuss the visualization of astrophysical data. Light properties of sunlight scattered by the atmosphere of Venus were measured by the Pioneer Venus Orbiter. One of the objectives of this mission was to determine the properties of the clouds and haze in the atmosphere.
Given the amount and complexity of the data, it is important to be able to browse through the data and select maps with interesting features. We built a system that reads the raw data, prepares it and extracts cloud features. The feature extraction is achieved by the following steps: selection, clustering, attribute calculation and iconic mapping. After data exploration a number of consecutive images with coherent moving cloud features, is found. From the center position and the time between two frames, a qualitative measure for the cloud velocities is derived. The obtained velocities are well in correspondence with generally accepted results.
Thus we have showed that visualization techniques are powerful tools to browse through the data, recognize cloud features and determine the motions of the features in time.

Keywords: scientific visualization, feature extraction, data exploration, astrophysical data.

1   Introduction

On May 20th 1978 the Pioneer Venus Orbiter was launched by a Centaur launch vehicle from Kennedy Space Center, Florida USA. It arrived at Venus the 4th of December 1978 and was placed in orbit, where it stayed for almost twelve years. Among the 17 instruments aboard was the Orbiter Cloud Photo Polarimeter, or OCPP. The instrument measured the intensity and polarization of sunlight reflected by the atmosphere. Unpolarized sunlight is absorbed and scattered by cloud particles, the intensity and polarization characteristics of the reflected light depend on the structure of the atmosphere. The principal objective of the Pioneer investigation was to determine the properties of the clouds and haze, including the vertical and horizontal distribution of the particles, cloud particle size and refractive index. Also the variations in cloud morphology and the nature of the cloud motions are investigated.

The 2Gb of data is organized in 4000 so-called maps that contain collections of 2D spherical coordinates (the radius is assumed to be fixed) and the measured quantities of the scattered light. Considering the large amount and the complexity of the data, visualization techniques are worthwhile. We developed a system to browse through the data, extract interesting features and visualize them. The combination of visual inspection and automatic feature extraction is a powerful tool to explore the data. The features extracted from the OCPP data are the cloud formations seen in the intensity and in the degree of polarization. For each cloud feature, a set of attributes is calculated and the features are visualized by ellipse icons. Displaying the ellipses of successive images gives us means to find coherent motion of cloud features. From the values of the center points in the attribute set and the time between two frames, we can give a qualitative measure for the cloud velocities. The obtained wind speeds are verified by results from astrophysical research.

These results are obtained in various ways (Rossow et al 1980 [3], Rossow et al 1990 [4], Toigo et al 1994 [7] and Smith and Gierasch 1996 [6]). The obtained velocities are divided in a zonal and a poleward component. It is generally accepted that the mean zonal wind speed corresponds to a solid rotation with an average equatorial speed of 93 m/s and the mean meridional wind is poleward, in the order of 10 m/s. This is proportional to an average global circulation periodicity of 4.5 days. The results are time dependent; they change over a period of weeks and they are a function of the latitude. These results will be used to verify the results obtained by our method.

This paper is organized as follows. In section 3 we will discuss the properties and some of the preprocessing of the OCPP data. Then we explain the feature extraction techniques in section 3, and in section 4 we present the results. Finally in section 5 the conclusions and future research topics are given.

2 The Pioneer Venus OCPP Data

2.1 Data generation

The orbiter, rotating around his axis, measured the data in scan lines across the surface, see Figure 1a. During every scan line one of the four different wavelengths (270, 365, 550 and 935 nm) is measured. The surface is thus scanned by about 50 scanlines for every wavelength. One planet surface measured in this way is called a *map*. The measured properties of the reflected light are: the intensity I, degree of linear polarization $|P|$ and the direction of polarization χ relative to the local scattering plane. Furthermore, the positions of the scan points, the sub-solar point and the sub-orbiter point are given in spherical coordinates. The coordinate system is fixed in space, with the positive x-axis pointing in the direction of the constellation Aries, i.e. it does not rotate with Venus.

wavelength	270	365	550	935
data	I	$|P|$	χ	
position	longitude		latitude	
scattering properties	μ_0	μ	θ	$\phi - \phi_0$

Table 1: Overview of the data measured by the Pioneer Venus OCPP.

In Figure 1b) a diagram of the situation is given as seen from the orbiter's point of view. In this Figure O is the sub orbiter point and S is the sub solar point. The visible hemisphere has as its center the O point, while the illuminated hemisphere has S as its center. These two hemispheres overlap in a symmetric segment on the planet's surface, bounded at one side by the day limb and at the other side by the terminator. The meridian which cuts it into equal halves is called the *symmetry meridian*. The line through O and S is called the *intensity equator*. Both are important symmetry axes [1].

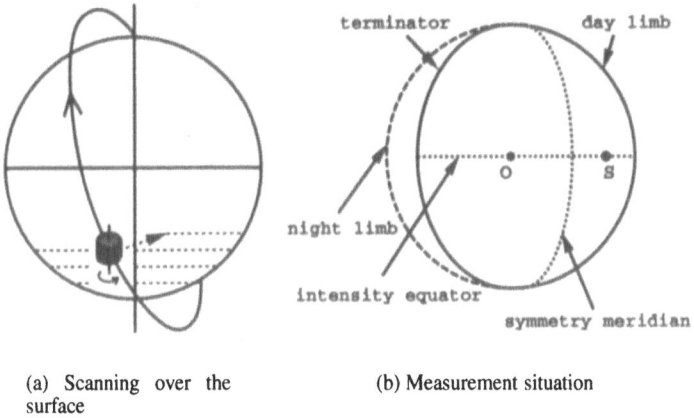

(a) Scanning over the surface

(b) Measurement situation

Figure 1: Diagram of Venus and the Pioneer Venus Orbiter.

With the positions of the sun and the orbiter, the local scattering geometry (Figure 2) is determined: μ_0 is the cosine of the angle between the incoming light and the normal, μ is the cosine of the angle between the outgoing light and the normal, the azimuth angle $\phi - \phi_0$ is the angle between the plane of incidence and the plane of reflection and the phase angle θ is the angle between the incoming and outgoing direction of the light. In Table 1 an overview of the data is given.

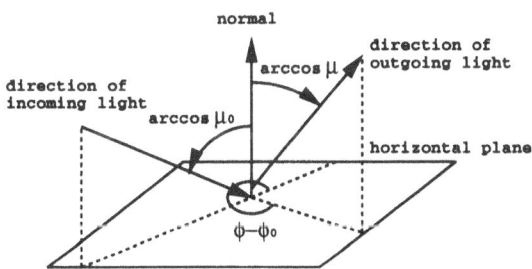

Figure 2: Local scattering geometry.

The scattering properties and the time between two measurements are different for each map. Since the polarization degree also depends on the phase angle, the data range in $|P|$ can be considerably different between two successive maps. Furthermore, the intensity and polarization quantities are not always related to each other. This causes the data to be inaccessible and hard to interpret. Therefore, it is important to be able to browse through the data: select specific map numbers, wavelengths and data quantities. Special routines to read the raw data were written in AVS. The visualization is realized by the following steps: resampling, limb darkening correction and data presentation.

2.2 Resampling

Since the data was measured in scan lines at non-uniform intervals, it is delivered as scattered data. In order to obtain the data on a regular grid, the data is resampled. This is achieved by a nearest-neighbour interpolation algorithm (Figure 3). The area around each grid point is divided into a number of sectors. For each sector the nearest raw data point, within a certain *search radius*, is determined. If for all sectors the nearest neighbour is found, the interpolation will take place by distance weighting over these points.

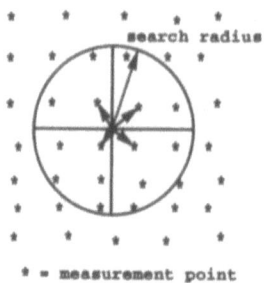

Figure 3: Resampling the raw data to a regular grid.

A good choice for the number of sectors is three or four. The search radius may not be too small: this means loss of data when the distance between raw data points is larger. However too large means searching in a large area, which means more computation time, while the effects are small. The best choice for the search radius is about two or three times the average distance between the raw data points.

2.3   Limb darkening correction

The geometry of a spherical object reflects the light with a certain *limb darkening*: the intensity is strongest in the middle and weaker near the edges. This is visible in the brightness distribution at wavelength $\lambda = 550$nm (Figure 4). For the higher wavelengths (550 and 935nm) scattering by gas molecules is less significant; therefore the influence of atmospheric properties on the intensity is small. However, the limb darkening also affects the lower wavelengths. This causes features near the limbs to be blurred or even become invisible.

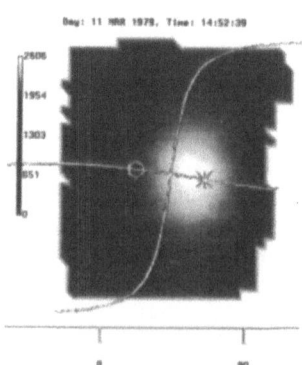

Figure 4: The limb darkening effect: the brightness distribution at wavelength $\lambda = 550$nm.

These features are enhanced by correction for the limb darkening based on Lambert's law. This law holds under the assumption that the light is reflected with equal intensity in all directions (perfectly diffuse reflection). This leads to the following correction procedure (Rossow et al [3]). First a disk integrated scaling factor I_d is calculated, then the appropriate zenith angle is subtracted from the raw intensity I^r.

$$I_d = \sum_n I_n^r [\sum_n \mu_{0n}]^{-1} \tag{1}$$

$$I_n^c = I_n^r - I_d \mu_{0n} \tag{2}$$

Thus the corrected intensities are an indication of intrinsic brightness deviations from the mean brightness over the disk. They can have positive or negative values, corresponding to relatively bright or dark features. This correction can be performed on the raw intensity data (e.g. before the resampling step). Two other correction methods have also been implemented; however, these do not give significantly better results [2]. Errors are estimated by calculating the Relative Difference Measure or RDM:

$$RDM = \sqrt{\frac{\sum_n (I_n^r - I_n^c)^2}{\sum_n (I_n^c)^2}} \tag{3}$$

where I_n^r is the raw intensity field in node n and I_n^c the correction field in node n. The accuracy of a method is measured by calculating the RDM between the brightness distribution at wavelength $\lambda = 550$nm and the correction field.

2.4 Data presentation

A colour presentation of the data can be given in two ways. Since the data is available in spherical coordinates, with the radius constant, it can be presented as a 2D map. The range of longitude is from -180 to 180 degrees and the latitude lies between -90 to 90 degrees. However, the data will be horizontally stretched out near the poles, which also deforms feature patterns. This can be avoided by projecting the data on a sphere (in Cartesian coordinates). In Figure 5 both presentations are displayed. The colours indicate the values of the scalar data. The time and date of the measurement are shown in the header. The positions of the sub orbiter and sub solar point are marked by o and * respectively, and the two symmetry axes are visualized by lines.

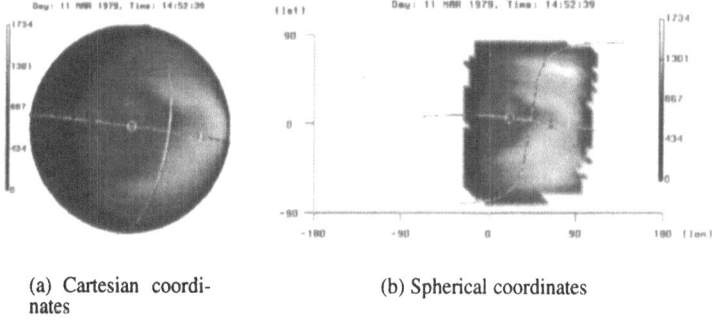

   (a) Cartesian coordinates   (b) Spherical coordinates

Figure 5: Two types of data presentation, day 169, $\lambda = 365$ nm.

Thus we have created means to browse through the data, select a specific map, wavelength and data quantity and view it with user defined colour maps in one of the projections shown in Figure 5. The browser can be used to search significant patterns or features in the data by visual inspection. These features can than be detected automatically and visualized as described in the next section.

3 Feature Extraction

In this section we will discuss our technique of feature extraction. This method was described earlier by van Walsum [8]. The following steps are performed:

- Selection.
  The creation of selections is achieved by evaluation of a selection expression. The expression is specified by a language that consists of many operators and functions (unary, binary, statistical and gradient functions). The language uses a C-like syntax. With this, one can create virtually any selection that is based on combinations of different data values, or quantities derived from data values, and on the grid node positions.

- Clustering.
  In order to handle regions of selected nodes as entities it is necessary to perform clustering. Neighbouring selected nodes are grouped into regions based on a connectivity criterion. Each group is labeled for reference.

- Attribute calculation.
  For each feature an attribute set is calculated in order to characterize the shape, size, position and orientation of the feature. Integration over the selected nodes leads to the volume/surface area, center point and the second moments.

- Iconic visualization.
  The attribute set is mapped onto the parameters of a geometric object or icon (van Walsum et al [9]). The goal of iconic mapping is to visualize essential elements of a data set in a symbolic way. The icon must be clear, compact, meaningful, and it should be related to the physical concepts and visual languages of an application.

This is a process of *abstraction*: a region in a data set is viewed and treated as an entity, and is visualized at a higher level. The meaning of a selection is changed: previously a selection was viewed as a set of nodes that contains interesting data, while here a selection relates to a feature. Therefore the term feature applies to any region for which an abstract representation and visualization is worthwhile. That is why the process of selection, clustering, and attribute calculation is called feature extraction.

3.1 OCPP cloud features

In order to extract cloud features from the OCPP data, first a suitable selection criterion has to be found. The selection expression works with threshold values, and thus it is important to have a mechanism to determine a suitable value. It may consist of an absolute value or a function of certain statistical parameters. It must be invariant for different maps. Finding a proper threshold is a matter of exploration of the data by the user. The definition of features cannot be given by universal rules, but depends on the underlying physics.

Generally speaking, a cloud feature is defined as a deviation from the average value. A region with values below the surrounding is a "dark" feature and a region with higher values is a "light" feature. Obviously statistical parameters such as the mean or the standard deviation (SD) are useful. If the distribution of the data is Gaussian, two thirds of the data values lie within one SD and 95% lies within two SD's from the mean. In both cases the criterion is invariant for different maps. Here another advantage of the limb darkening correction arises: the correction creates a Gaussian distribution. The relatively dark features have negative values and the bright features have positive values. In the expression below a selection of dark cloud features from the corrected intensity field at wavelength 365nm, labeled **I365**, is made.

$$\begin{aligned}
\textbf{ranged} &= (\ \textbf{I365} - \text{avg}(\textbf{I365})\ )\ /\ \text{sd}(\textbf{I365}) \\
\text{threshold} &= 1.0 \\
\textbf{select} &= \textbf{ranged} < \text{-threshold}
\end{aligned}$$

The simplest way to visualize the position and the size of a selected feature is with an ellipse icon. The average position of the cluster is mapped to the position and the second moments determine the length and the orientation of the axes. In case of the 2D ellipse a minimal attribute set of five parameters is needed. The surface area, center point and second moments are calculated by means of integration over the selected nodes. In Figure 6 a selection is made from intensity data with a wavelength $\lambda = 365$nm. The intensity data is corrected and the selection is made by the expression described above, e.g. dark features are extracted. The selected points are marked by the small white crossmarks and the ellipses visualize the cloud features.

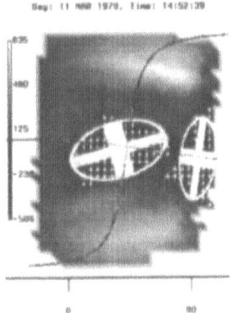

Figure 6: Iconic presentation of the selected features.

3.2   Cloud tracking

Now we have means to automatically extract features from the OCPP data. In order to find a series of maps with coherent moving features, we first make a rough selection from all maps. The following two restrictions are applied: the time between two successive images must be less than 12 hours and the phase angle θ must be less than 80°. The first restriction is based on the expected speed of motion (4.5 days circulation) and the size of the visible part. Half the planet's surface is visible and this part becomes smaller with increasing phase angle; therefore the second criterion is introduced. These two criteria reduce the number of useful maps to a small number of series.

The remaining maps are then explored for coherent features using the browser. Intensity and polarization data may show different features, so both types of data are explored. Correspondence between two cloud features in two frames is established manually, based on the following criteria:

- Position.
  The feature in the second frame must be at the position expected from the position and velocity (known in advance) of the feature in the first frame.

- Volume (or surface area).
  The two features must have approximately the same size.

From the calculated center positions and the time interval between two frames, the velocity of the features can be estimated.

4 Results

4.1 Cloud tracking

The series shown in Figure 7 is an example of a series with coherent moving features. It covers about 28 hours and the phase angle changes between $63.2 < \theta < 68.8$. Since the phase angle is almost constant, the threshold can be taken as an absolute value. The selection is based on the polarization degree: $|P| > 3.0$. In the figure six frames are shown in which features move from right to left. In the first three frames two features are visible. Based on the positions and consistency we can say that the left feature moves out of the image and that only the right feature is still apparent in the fourth frame. In the frame f) it also almost disappears.

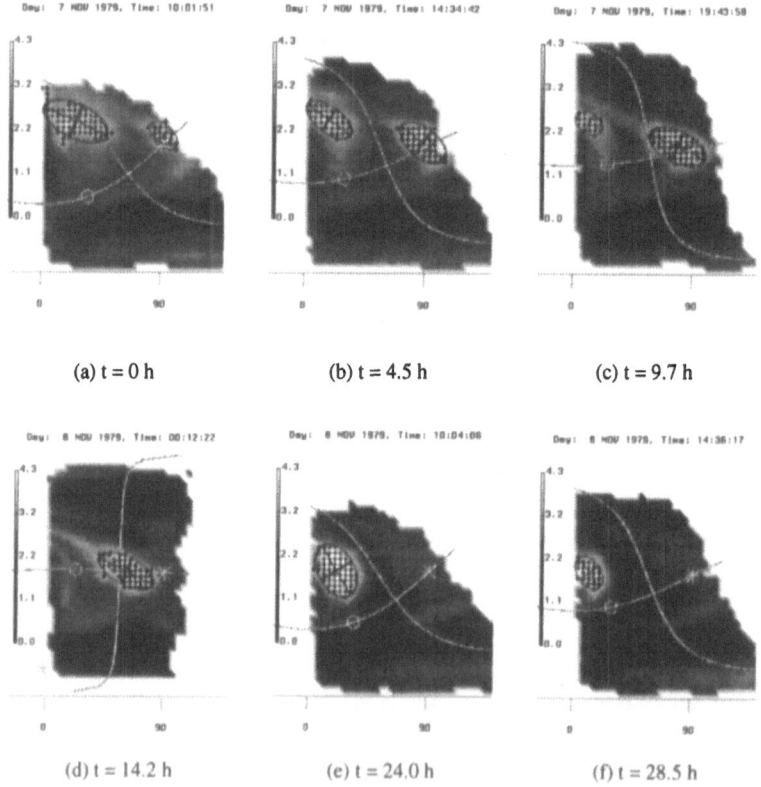

Figure 7: Determination of circulation velocity by means of feature tracking, day 507-512, $|P|$ data, $\lambda = 365$ nm.

In Figure 8 the longitude and the latitude of the second feature are plotted as a function of the time. In about 28.5 hours the feature has moved from longitude = 91.9 to 12.5 degrees. During this period the latitude was almost constant. This means that the circulation time is about 5.4 days. However in the beginning and at the end, part of the feature was outside the measurement area. This affects the position of the center point and the approximation of the circulation velocity. The effects of the borders may be avoided by checking the surface area of the feature: the surface of the feature in the first and the last frame are much smaller (±300 degrees^2) than the rest (±750 degrees^2), which is a reason to skip them. The second is also near the border, however it is almost as large as the others (±680 degrees^2) and therefore we can take it into account. The result is a circulation time of 4.4 days. This result is very well in correspondence with the expected average of 4.5 days. This supports the conclusion that the extracted feature is a real moving cloud formation.

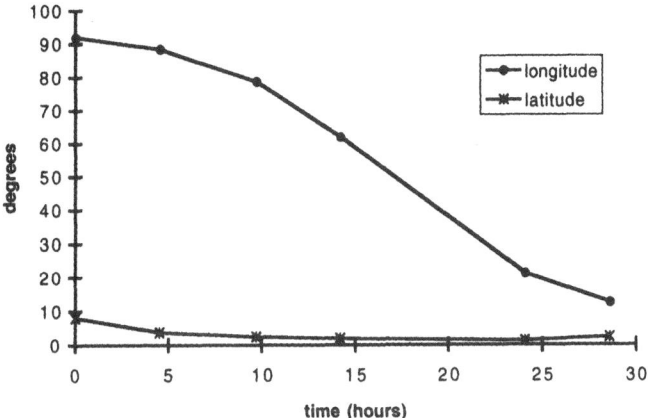

Figure 8: Movement of the center point of the feature.

5   Conclusions and Future Research

The data measured by the Pioneer OCPP experiment consists of a large, complex data set. The scattering geometries and the time between two measurements differs per map. This causes the data to be very inaccessible and hard to interpret. Therefore it is important to be able to browse through the data and select maps with interesting features. We built a system that reads the raw data, prepares it and extracts cloud features automatically. The correction of the intensity data for the limb darkening gives a good enhancement of features. A better distinction can be made between dark and light features.

Features are tracked by performing the feature extraction on consecutive images. However it takes some data exploration to find a suitable series of images with coherent moving features. The center positions of these features are used to approximate the velocity. Features near the border might influence the velocity, but the surface area can be used as an indication to include them or not. Tracking the large scale cloud features in the Pioneer Venus OCPP data results in a circulation time of 4.4 days. This is in correspondence with the average 4.5 days rotation found by Rossow at al [3].

We believe that these results justify the development of an algorithm for automatic feature tracking based on calculated attribute sets. The attribute set should hold the proper

94

information in order to determine feature translation, rotation and scaling. In addition, the algorithm should be able to recognize the events like: continuation, creation, dissipation, bifurcation and amalgamation [5]. In order to recognize these events, a number of test criteria for the attributes can be implemented.

Ellipses do not always give a good impression of size and position. For instance a long curved feature is not properly visualized by an ellipse. In order to visualize such features, icons must be used which also account for the shape of the feature. One way to generate shape attributes, is the determination of the medial axes or skeleton. This may be achieved by morphological operations on a cluster of selected nodes.

6   Acknowledgments

The authors wish to acknowledge the stimulating discussions with dr. J.F. de Haan and drs. R. Braak on the interpretation of the OCPP data.
This work is supported by the Netherlands Computer Science Research Foundation (SION), with financial support of the Netherlands Organization for Scientific Research.

References

[1] J.W. Hovenier. Principles of symmetry for polarization studies of planets. *Astron. & Astrophys.*, 7(1):86–91, 1970.

[2] Freek Reinders. Visualization of the pioneer venus ocpp data. Technical report, Delft University of Technology, The Netherlands, 1997.

[3] William B. Rossow, Anthony D. Del Genio, and Timothy Eichler. Cloud-tracked winds from pioneer venus ocpp images. *J. of Atmos. Sci.*, 47(17):2053–2082, 1990.

[4] William B. Rossow, Anthony D. Del Genio, Sanjay S. Limaye, Larry D. Travis, and Peter H. Stone. Cloud morphology and motions from pioneer venus images. *J. of Geophys. Res.*, 85(13):8107–8128, 1980.

[5] Deborah Silver and X. Wang. Volume tracking. *Proc. IEEE Visualization '96*, pages 157–164, 1996.

[6] Michael D. Smith and Peter J. Gierasch. Global-scale winds at the venus cloud-top inferred from cloud streak orientation. *Icarus*, 123:313–323, 1996.

[7] A.D. Toigo, P.J. Gierasch, , and M.D. Smith. High-resolution cloud feature tracking on venus by galileo. *Icarus*, 109:318–336, 1994.

[8] Theo van Walsum. Selective visualization on curvilinear grids. *PhD Thesis*, 1995. Delft University of Technology, The Netherlands.

[9] Theo van Walsum, Frits H. Post, Deborah Silver, and Frank J. Post. Feature extraction and iconic visualization. *IEEE Trans. on Visualization and Computer Graphics*, 2(2):111–119, 1996.

Editors' Note: see Appendix, p. 180 for colored figures of this paper

Visualization in topology: assembling the projective plane

Stanislav Klimenko[1], Igor Nikitin[1]
Martin Göbel[2], Henrik Tramberend[2]

[1] Russian Center of Computing for Physics and Technology
RCCPT, Protvino, Moscow region, 142284, Russia
Klimenko@mx.ihep.su
[2] Visualization and Media Systems Design – VMSD,
German National Research Center for Information Technology
GMD, Schloss Birlinghoven, 53574 Sankt Augustin, Germany
Martin.Goebel@gmd.de

Abstract. Assembling the projective plane in 3D space from a Möbius band and a disc is animated. Various properties and representations of the projective plane are visualized.

Introduction

Two different objects which can be continuously 1-to-1 mapped one to another are equivalent from topological point of view. They can be considered as different models of one topological object. The projective plane is one of these archetypal objects. It is particularly interesting, because it is most simple among topologically non-trivial objects and it is one of the elementary blocks, from which more complicated objects are constructed.

By definition, the projective plane is a set of all straight lines in 3D space, passing through the origin. Each line is uniquely defined by its intersection with the sphere around the origin, but two opposite points on the sphere define the same line. Therefore, another model of the projective plane is a sphere in 3D space, on which the pairs of opposite points are identified, i.e. considered as one point. We can always imagine that this sphere is mapped somewhere, e.g. to 3D space again, using a mapping, which actually glues the opposite points in one. This will provide us with model of the projective plane as a surface in 3D space. Also, we can use only a semi-sphere to represent the projective plane, and take into account that opposite points on its equator should be identified. Topologically the semi-sphere is equivalent to a disc (or rectangle), so the projective plane can be defined as a result of patching the opposite points on the disc.

The projective plane and its generalizations are the top objects under research in different areas of science. These objects naturally appear in physics, in those problems, where some axis defines the state of the system, and the replacement of an axis direction with its opposite does not change the state. Such identification of opposite directions usually changes the deepest properties of the system. One can see this in a remarkable video of T.Delmarcelle and

L.Hesselink [1], devoted to topology of tensor fields. This video clearly shows, how the replacement of a vector field with the field of non-directed axes (eigen axes of the tensor field) creates stable singular points, not typical for vector fields. Such singularities exhibit themselves, particularly, as observable defects in liquid crystals [2], which consist of long non-directed molecules, tending to line up parallel their axes.

In this work we visually present two models of the projective plane in 3D space and animate a solution of a famous topological problem, which being said with the words of G.K.Francis, was a gate to topology for a number of generations of the students [3].

1 Patching Möbius band and disc

Let's consider a Möbius band and a disc. The Möbius band has one side and *one edge*. The disc has also one edge, so we can patch them together along the edges. We can do this as shown on fig.1 (see color plates). We cut the disc along the radius, twist it twice, glue the cut and put down this surface onto the Möbius band. In this their edges coincide, and we patch them. The obtained surface is called *cross cap* (fig.2). It is one possible image of the projective plane in 3D space.

This surface has many interesting properties. Particularly, it is not orientable, like the Klein bottle or the Möbius band contained in it. If we consider a small object on the surface and move it along a definite closed path, we will obtain the object with reverse orientation. Therefore, we cannot specify global orientation on the surface. We can easily show this property, using the animation (fig.3).

The cross cap is the most simple but not quite satisfactory model of the projective plane. The best model (called *embedding*) should not have self-intersections. However, in 3D space this model is impossible due to a general topological theorem, which prohibits embedding of closed non-orientable surfaces into 3D space. Therefore, we should consent with self-intersections, keeping in mind, that each point on self-intersection represents two points of the projective plane.

Further, we see that the line of self-intersection on the cross cap terminates in two points, called *points of pinch*. The surface near these points has a complicated structure. Particularly, it is not smooth here: there is no tangent plane to the surface in the pinch point, there is no normal vector as well. When we approach the pinch point from different directions, we will achieve different limiting positions for tangent planes and normals[3], see fig.4. Namely for this reason

---

[3] Note, that limits of unit normals lie in a plane and cover the big circle on unit sphere. This explains one finer property of the pinch point. Let's consider some projection of the surface onto a plane. The point of the surface is mapped onto contour of the projection, if the normal in it is orthogonal to the direction of projection. For any direction of projection, a normal orthogonal to it exists on limiting circle (because the set of all unit normals, orthogonal to a fixed direction, is another big circle, and two big circles on the sphere always have intersections). Therefore, the pinch point is always projected onto a contour of the surface (visible or hidden by other parts).

this image of the projective plane does not satisfy topologicists. Again, due to a general theorem, there exists a smooth mapping *(immersion)* of any 2D surface into 3D space, with self-intersection but without singular points, looking like the image of the Klein bottle (fig.3). The immersion of the projective plane in 3D space was constructed by Werner Boy in 1900, but its explicit analytical representation was found by Francois Apery just in 1987. This surface and continuous deformation of cross cap to it are shown on fig.5. The phenomena of pinch points creation and cancellation can be found on intermediate stages.

The Boy's surface has the line of self-intersection organized in three loops linked in one point (node). Actually, it is a single smooth closed connected curve (topological circle), which intersects itself in this node. The node is a triple point of the surface, in its vicinity the surface looks like three intersecting planes.

2 Visualization methods

The described surfaces and their deformations are shown by C++ application, realized on SGI platform. Figures placed in this article are taken from the movie, generated by this application. The position of view point and various parameters of the surfaces can be changed interactively during the movie. Five minutes video is prepared using the application. Our final intent is the implementation of this application in the system of virtual environment Cyberstage (see Appendix).

In this Section we will describe in details the methods and tools of visualization under the use.

Parametrization of key shapes. The cross cap is generated by rotating ellipse with variable height:

$$\mathbf{r}_c(\theta, \phi) = \begin{pmatrix} (1 + \cos 2\theta) \cos 2\phi \\ (1 + \cos 2\theta) \sin 2\phi \\ \sin 2\theta \, h(\phi) \end{pmatrix}, \quad h(\phi) = \sin \phi. \tag{1}$$

Let's compare this surface with the sphere:

$$\mathbf{r}_s(\theta, \phi) = (\cos \theta \cos \phi, \ \cos \theta \sin \phi, \ \sin \theta). \tag{2}$$

The shift $\phi \to \phi + 2\pi$ conserves the point on both surfaces. The poles of the sphere correspond to lines $\theta = \pm \pi/2$ on the parameter plane. The mapping (1) glues these lines to a point again. Therefore, one can consider (1) as a continuous mapping of the sphere into 3D space. Further, the transformation

$$\theta \to -\theta, \ \phi \to \phi + \pi \tag{3}$$

interchanges the opposite points on the sphere, but conserves the point on cross cap[4]. Therefore, the mapping (1) identifies opposite points on the sphere, and really represents the projective plane. The map of the sphere – a square $-\pi/2 <$

---

[4] Note, that this transformation reverses the orientation of parameters plane and also the normal $\mathbf{n} = \partial \mathbf{r}/\partial \theta \times \partial \mathbf{r}/\partial \phi$. Let some object continuously moves on parameters

$\theta \leq \pi/2$, $0 \leq \phi < 2\pi$ covers the cross cap twice. A square $-\pi/2 < \theta \leq \pi/2$, $0 \leq \phi < \pi$ covers it once. The line of self-intersection corresponds to $\phi = 0$, the pinch points are located at $(\theta, \phi) = (0,0)$ and $\theta = \pi/2$. The band $S(-1 + \cos\phi) < \theta < S(1 + \cos\phi)$ is the Möbius band, contained in the cross cap, and the part outside this band is the twisted disc. Here $S < \pi/4$, and we have fixed $S = 0.4$.

Apery parametrization is used to represent the Boy's surface (see [3]):

$$\mathbf{r}_B(\theta, \phi) = A \begin{pmatrix} r_1 \cos 2\phi \\ r_1 \sin 2\phi \\ 1 \end{pmatrix} + B \begin{pmatrix} r_2 \cos \phi \\ -r_2 \sin \phi \\ 0 \end{pmatrix}, \tag{4}$$

$$A = \frac{\cos^2 \theta}{1 - \beta \sin 2\theta}, \quad B = \frac{\sin \theta \cos \theta}{1 - \beta \sin 2\theta}, \quad \beta = b \sin 3\phi,$$

and $r_1 = \sqrt{2}/3$, $r_2 = 2/3$. At $1/\sqrt{6} < b < 1$ there are no pinch points on the surface. We have preferred $b = 3/(2\sqrt{6})$. As for the cross cap, lines $\theta = \pm\pi/2$ are mapped to the origin and the transformation (3) conserves the point on the surface. Therefore, (4) is the projective plane also.

The Klein bottle, shown on fig.3, was parameterized as follows:

$$\mathbf{r}_K(u, v) = (R_x, R_y, R_z^2), \tag{5}$$

$$\mathbf{R}(u, v) = \mathbf{R}_0(u) + \rho(u)(\mathbf{e}_1(u) \cos v + \mathbf{e}_2 \sin v),$$

$$\mathbf{R}_0 = \begin{pmatrix} a \sin 2u \\ 0 \\ b \cos u \end{pmatrix}, \quad \mathbf{e}_1 = \begin{pmatrix} b \sin u \\ 0 \\ 2a \cos 2u \end{pmatrix} \frac{1}{|\mathbf{R}_0'|}, \quad \mathbf{e}_2 = \begin{pmatrix} 0 \\ 1 \\ 0 \end{pmatrix},$$

$\mathbf{e}_1$ and $\mathbf{e}_2$ are unit vectors, orthogonal to $\mathbf{R}_0'$,

$$\rho = \rho_0 + \rho_1 \sin^{2n}(u - u_0),$$

with the following values of parameters:

$$a = 1/4, \ b = 1, \ \rho_0 = 0.1, \ \rho_1 = 0.6, \ u_0 = 0.3, \ n = 4.$$

Transformations $v \to v + 2\pi$ and $(u \to u + \pi, \ v \to \pi - v)$ conserve the point on the surface, and the square $0 \leq u < \pi$, $0 \leq v < 2\pi$ covers the surface once.

This parametrization is constructed using the following idea (fig.6). Let's take S-shaped tube, then thicken it (modulating its radius) and reflect the bottom part to the top, replacing $z \to z^2$.

For the animation of deformations we cut the parts of these surfaces (e.g. for the disc twisting) and use linear homotopies between them, like $\mathbf{r} = (1 - t)\,\mathbf{r}_c + t\,\mathbf{r}_B$, $0 \leq t \leq 1$.

---

plane between the points, connected by this discrete transformation. Locally the surface has two sides, and if the object moves "always on one side", it will arrive to the same point of cross cap, but from another side. It will be reversed in the sense, that 2D habitant of the projective plane, living in the vicinity of this point, will see the returning object mirror-reflected.

Fig.6. Construction of Klein bottle.

Tools. Open Inventor [4] is used as the main tool for visualization. It is well suited for animation of complicated 3D shapes. It includes the viewers, allowing user to change interactively the position of the viewpoint, and convenient system of event callbacks to user's procedures, which can perform any actions in the scene (e.g. set the current time and time step for the movie, switch different options of viewing, record the position of a camera etc.).

Transparent surfaces. Open Inventor supports the realistic rendering of transparent surfaces, but only for simple shapes (sphere, cone, cube). The rendering of arbitrary shapes, represented e.g. as triangle strip sets, requires preliminary sorting of triangles according their position in z direction. This creates serious problems:

- sorting is time-consuming procedure;
- intersecting triangles can not be correctly sorted, due to this fact the lines of self-intersection on the surface become non-smooth. In principle, one can determine the position of these lines and subdivide the triangles along them, in order they will not have intersections. However, this task is not easy for real-time calculations, if the surface moves.

To avoid these problems, we have used the following method for *simulation* of transparency. For each object we create its small transparent copy near the viewpoint[5]:

$$\mathbf{r}_{copy} = \mathbf{r}_{camera} + (\mathbf{r}_{original} - \mathbf{r}_{camera})/W. \tag{6}$$

After the projection, the copy exactly overlays the original – we look on the opaque object through its transparent copy and therefore see its hidden parts. These parts are visible only on the copy, and therefore are less bright than front parts. By changing the transparency of the copy, we can influence the seeming

---

[5] Whenever the camera position is changed, the position of the copy should be recalculated. For stereo viewing two copies are needed, own for each camera. The weight W should be great to avoid the intersection of the copy and original. The expression is written for the perspective camera. For orthographic camera one can write analogous expression.

transparency of the image (when the copy is absolutely transparent, the object seems absolutely opaque). The copy is rendered using a simple additive algorithm (see [4]), which does not distinguish front and back parts. However, on the final image these parts can be distinguished. Note, that some triangles forming the surface have intersections, however the described method gives smooth intersection lines (see figures).

NOTES.

1. In this method the rendering of hidden surfaces is not quite correct. Namely, if we have two surfaces, hidden by a third one, we can not distinguish front and back parts of *hidden* surfaces (and as a result, can not see hidden lines of self-intersection). However, in the cases when only the presence of the hidden structure should be shown (but not its realistic rendering), this method works well. Because this method slows the rendering only by half, we are able to perform the rendering for the movie in real time.

2. We will describe an idea, how one can perform fast realistic rendering of intersecting transparent surfaces. We should modify the common z-buffer procedure: let's store z-coordinates not only for current pixel, coming to given point on xy-plane, but for *all* pixels placed in this point, as a list ordered respective to z. Also let's store their colors c and transparencies t ($t = 0$ - solid, $t = 1$ - absolutely transparent). When all pixels are put to this buffer, let's calculate the resultant color using the formula

$$\left.\begin{array}{llll} z_0 > z_1 > z_2 > \dots \\ c_0 \quad c_1 \quad c_2 \quad \dots \\ t_0 \quad t_1 \quad t_2 \quad \dots \end{array}\right\} \quad c = (c_0 + t_0 \cdot (c_1 + t_1 \cdot (\dots))) \cdot T,$$

here each pixel suppresses the colors of all pixels lying below by the coefficient t. Common coefficient T should be chosen to constrain $c_i \leq 1$.

Using this method, we will not encounter the problem with incorrect rendering of intersecting triangles, because the sorting now is performed at the level of pixels. Also, when the length of the pixel's lists (number of layers of the surface) is not great, their sorting will be not much slower than ordinary comparisons in the standard z-buffer.

By our opinion, the test and implementation of this algorithm will be of great use.

Colors and textures. Colors are used to emphasize the lines of self-intersection on the surface[6]. The textures play the same role. Also, the smooth surfaces should be textured for stability of the stereo image in virtual environment.

There is one problem with mapping colors and textures onto topologically non-trivial object without breaks of continuity and smoothness. For the projective plane the color vector should be a smooth function on the parameter plane: $c_i(\theta, \phi)$, $i = r, g, b$, conserving in transformation (3). Additionally, it should be constant on the lines $\theta = \pm\pi/2$, which are mapped in one point. The mapping from (θ, ϕ) to coordinates (x, y) inside some texture image should have the same properties. The following function with these properties was chosen: $f(\theta, \phi) = (1 + \sin 2\theta \, \cos(\phi - \phi_0))/2$ and was used with $\phi_0 = 0$ for c_r, with $\phi_0 = 2\pi/3$ for c_g, with $\phi_0 = 4\pi/3$ for c_b; with $\phi_0 = 0$ for x and with $\phi_0 = \pi/2$ for y.

---

[6] Or, if one likes, encode additional coordinates in higher dimensional "coordinate × color" space, in which the surface is *embedded*.

NOTES.

1. Textures are ordinary prepared in such way, that they can be periodically continued in x and y directions, actually, they are textures for a torus. If they are "constant" on top and bottom, they are textures for a sphere. However, 1-to-1 mapping $(\theta, \phi) \leftrightarrow (x, y)$ of the projective plane to a plane, torus or sphere *does not exist* (otherwise, they will be topologically equivalent). The constructed mapping $(\theta, \phi) \rightarrow (x, y)$ is not invertible, two (or more) points (θ, ϕ) will be mapped in the same point (x, y). Some parts of the texture have duplicates on the projective plane (fig.7).

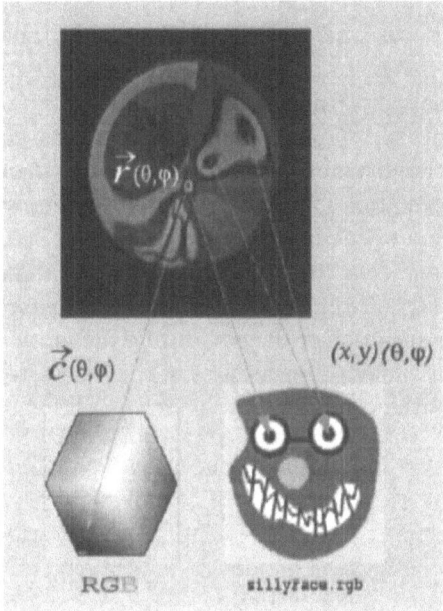

Fig.7. Coloring and texturing is performed by smooth mappings $c(\theta, \phi)$ and $(x, y)(\theta, \phi)$ of the surface in color space and in texture plane respectively. The mapping $(x, y)(\theta, \phi)$ cannot be invertible due to the difference in topology of the surface and the texture plane. As a result, some parts of the texture have duplicates on the surface.

Namely here the topology of the object exhibits itself. If we intend (for some purpose) to map the texture 1-to-1 on the object with handles, we should use very special textures. And a texture with some group of symmetry is actually the texture for some topologically non-trivial object (orbifold[7]).

The 1-to-1 texture for the projective plane should have "shift-reflectional" symmetry (3). So, we should either find (or prepare) such texture or use the described method to create smooth but not 1-to-1 texture from arbitrary one. We've used the second possibility.

---

[7] see http://www.ornl.gov/ortep/topology/orbfld1.html

2. All considered surfaces are embedded into 6D space "3 coordinate × 3 colors", they are smooth (as one can easily check) and have no self-intersections there (intersecting parts have different colors). Actually we use 2 color dimensions in 3D color space: a plane $c_r + c_g + c_b = Const$. What will happen, if we will use only one color dimension (e.g. define c_r as before, but $c_g = 1 - c_r$, $c_b = 0$) and consider 4D "coordinate × color" space? The cross cap will remain to be embedded in this space. But the Boy's surface will not. Due to a definite topological obstacle [3] the Boy's surface cannot be elevated to embedding in 4D space. If we use the colors from some *linear* palette (rainbow), there will always be a point on the loop of self-intersection, where the intersecting parts have the same color. This is the self-intersection (double point) of the surface in 4D space. To eliminate this point, we should separate the intersecting parts in one more color dimension.

The image of the fish was kindly presented to us by Natalia Lemesheva. We have transferred it from her picture onto the projective plane, scanning the image in SGI rgb+α format and using it as the transparent texture on a separate moving triangle strip set. This triangle strip set was constructed as follows. We triangulate a rectangle, moving on the parameters plane between the points, connected by transformation (3). Then we map it on the surface and slightly shift it in positive direction of the normal at each vertex. The area of the rectangle is changed in mapping from the parameters plane to the surface, and the ratio of these areas depends on the position in the parameter plane. Therefore the fish suffers the great variation of the size during the motion. To compensate this effect, we permanently change linear sizes of the rectangle on parameters plane, multiplying them by the coefficient

$$ C = Const \sqrt{\frac{S_{\text{plane}}}{S_{\text{surf.}}}} = \frac{Const}{\sqrt{\left| \frac{\partial \mathbf{r}}{\partial \theta} \times \frac{\partial \mathbf{r}}{\partial \phi} \right|}} \; , \tag{7} $$

evaluated at the central point of the rectangle. We did not try to compensate other distortions of the image.

Additionally, we connected to the fish a camera with wide view angle and displayed the view, which "the fish sees", in a separate window.

3 Related works

There are a number of works devoted to visualization of the projective plane. The topology of the projective plane is considered in details in remarkable books by George Francis [3] and Anatoly Fomenko [5], different models of the projective plane and various methods of their construction are described there. A number of beautiful images of the projective plane and MPEG movies as well can be found in the following WWW sites:

* homepage of the Center for the Computation and Visualization of Geometric Structures at the University of Minnesota:

  `http://www.geom.umn.edu/zoo/toptype/pplane`

* dissertation on mathematical visualization by David Banks:

  `http://www.cs.msstate.edu/~banks/research/math.html`

* "Steiner surfaces" page by Adam Coffman:

  `http://www.math.uchicago.edu/~adamc/steinersurface.html`

4 Conclusion

As a conclusion, we summarize new features, involved in our visualization of projective plane:

- the process of patching of the disc to Möbius band is animated;
- to demonstrate the non-orientability of the surface and to show the structure of normal field near its singular points we animate the motion of small objects on the surface;
- to show self-intersections of the surface we use simultaneously colors and textures;
- to show the inner structure of the objects we use simultaneously the transparent surfaces, windows and additional viewport from the observator, moving inside the object.

Also we have developed and tested the simple methods for rendering of intersecting transparent surfaces and smooth mapping of color and texture onto topologically non-trivial objects. These methods can be applied in many challenging problems of scientific visualization. Particularly, we plan to use them for visualization of string dynamics (see [6] and fig.8), which creates the objects similar to described here. Incorporation and study of these objects in virtual environment is a subject of our future work.

Appendix: The Cyberstage

GMD's *Cyberstage* is CAVE-like installation, consisting of 3 rear projection walls (3.0m x 2.4m) and a ground projection (3m x 3m) with high quality video projectors. *Cyberstage* is installed in a 6m x 10 m room, requiring triple mirroring for the projection. Images are generated by a dual pipe SGI ONYX IR. Stereo viewing is provided with LCD shutter glasses. Head tracking is done with a magnetic space tracker. An acoustic floor has been integrated into the *Cyberstage* to display the sense of vibration to users. A loudspeaker-based surround sound system is completing the installation.

Acknowledgments

We are grateful to Irmtraud Fritz-Hassaïne, Klaus-Günter Rautenberg, Martin Suttrop, Wolfgang Vonolfen and Natalia Lemesheva for their valuable contribution in this work. Two of us (S.K. and I.N.) thank all people from Visualization and Media Systems Design department of GMD for their hospitality and friendship.

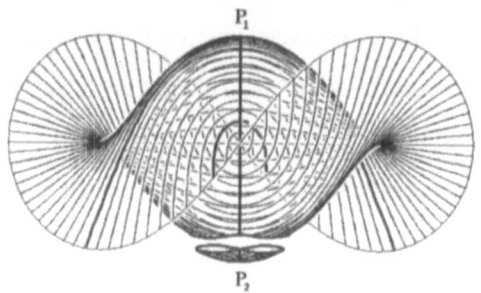

Fig.8. Theory of strings, physical theory, pretending to be the Theory of Everything, represents elementary particles as minimal surfaces, analogous to soap films, but placed in Minkowsky space-time. The minimal surfaces, correspondent to low level excitation of the particles, being projected from the space-time to the space, have a topology of *Sudanese surface*, the Möbius band with (almost) circular boundary. Sudanese surface, shown on this figure, is the main ingredient of standard visualization of projective plane assembling: to obtain the projective plane from Sudanese surface we should only patch its circular boundary by the disc.

References

1. T. Delmarcelle and L. Hesselink, "The Topology of Symmetric Second-order Tensor Fields", Proceedings IEEE Visualization '94, October 17-21, 1994, Washington, D.C., pp. 140-147. Material is available on WWW site
   `http://www.nas.nasa.gov/NAS/RelatedPapers/StanfordTensorFieldVis/home.html`

2. James P. Sethna "Order Parameters, Broken Symmetry, and Topology" in 1991 Lectures in Complex Systems, Eds. L. Nagel and D. Stein, Santa Fe Institute Studies in the Sciences of Complexity, Proc. Vol. XV, Addison-Wesley, 1992.
   `http://www.lassp.cornell.edu/sethna/OrderParameters/Intro.html`

3. George K. Francis, Topological Picturebook, Springer-Verlag, 1987.

4. Josie Wernecke, Open Inventor Architecture Group, The Inventor Mentor: Programming Object-Oriented 3D Graphics with Open Inventor, Release 2, Addison-Wesley, Reading, Massachusetts, 1994 (ISBN 0-201-62495-8);
   Josie Wernecke, Open Inventor Architecture Group, The Inventor Toolmaker: Extending Open Inventor, Release 2, Addison-Wesley, Reading, Massachusetts, 1994 (ISBN 0-201-62493-5).

5. A.T.Fomenko, Descriptive Geometry and Topology, Moscow University Press, 1992.

6. Klimenko S.V., Nikitin I.N., Talanov V.V. "Visualization of Complex Phenomena in String Theory", in Proc. Computer Animation'95, April 19-21, 1995, Geneva, Switzerland, pp.181-189.

Editors' Note: see Appendix, p. 181 for colored figures of this paper

Combining Wavelet Transform and Graph Theory for Feature Extraction and Visualization

Christoph Lürig, Roberto Grosso and Thomas Ertl

Computer Graphics Group, University of Erlangen
Am Weichselgarten 9, 91058 Erlangen, Germany

Abstract. In the process of visualizing 3D MRI or CT data using techniques such as isosurfacing or direct volume rendering, one is confronted with two problems. The first one is that there is no distinction between important and unimportant data. The second one is the difficulty to find a meaningful mapping of the measured scalar values to the graphical attributes used for the visualization. These problems are addressed by the special segmentation procedure presented in this paper. The basic idea is to apply graph algorithms to find important structures and to assign multidimensional information to these structures with the help of wavelets. This additional information can be used to generate graphical attributes for rendering. Several aspects emerge from the interaction of both theoretical concepts.

Introduction

One problem of visualizing medical data by means of volume rendering is, that simple transfer functions are often not sufficient enough to produce meaningful images. This is especially true for MRI data sets, where a measured value may correspond to several types of tissue. In this case the use of transfer functions produces irritating images, as there is little correspondence between the assigned color and the tissue. A first approach to solve this problem has been introduced by Levoy [6] who includes gradient information in the transfer functions. Westermann [10] integrates feature extraction by using the wavelet transformation to evaluate the spatial frequency of the visualized data.

In this paper we present a new method for segmentation and feature extraction for visualization purposes. The main idea is to find and to extract edges from the data set and to compute a feature vector, that may be used for segmentation or for the definition of transfer functions. We developed an algorithm that combines wavelet theory and graph theory, and we applied this to the two dimensional case. We also discuss how these concepts can be generalized to three dimensions.

Edge detection, line segmentation and scale space filtering especially with wavelets became popular and powerful techniques in image segmentation and feature extraction. The basic idea of scale space filtering was first described by Witkin [11]. A filter bank is applied to a given image, which extracts features at different levels of detail. This level of detail aspect was used by Bijaoui [2] to identify different astronomical structures. The filter coefficients for a given point in the image may be directly used as a feature vector as it was proposed by Leite and Hacock [5].

A filter technique, that is capable of scale space filtering, is the wavelet transformation [2, 5]. Many important results about wavelets and feature extraction have been published by Mallat et al. [7, 8, 9]. Berkner et al. [1] analyze the local maxima of the wavelet transformation and the evolution of the wavelet coefficients across different scales to obtain analytical properties of the underlying function which can be used for image segmentation in the medical context.

It can be shown, that the continuous wavelet transformation for a single scale is equivalent to the edge detection method of Canny [3]. Thus, a connection between wavelets and powerful edge segmentation techniques can be established. Several aspects emerge from the interaction of these theories. The Canny operator itself can only detect the maxima for each of the different scales. The wavelet theory provides a framework for obtaining further information, that can be derived from the interaction of the different scales. A commonly used edge processing step is the grouping of pixel maxima into lines for example using graph-theoretical methods as described by Zahn [12]. A different method for detecting strings is the snake method, which requires an initial segmentation of pixels the to drive the external forces. This differs from our approach, where the segmentation is done on the lines themselves. One snake is just capable of finding one closed structure. This differs from our approach, where multiple and also open structures may be segmented.

We present an algorithm, that is capable of identifying lines and of generating a feature vector for every line, that contains information about (1) the Euclidean length, (2) the Hölder exponent, which is a measure of smoothness, (3) the coarsest scale on which the line has a representation, (4) a strength value, that includes gradient information and (5) an unsharpness coefficient. In contrast to the approach followed by Hacock [5] the resulting vector does not vary in dimension with the scales involved but the more scales are used the more exact the resulting values will be. A major contribution of this paper is a new algorithm based on wavelets and graphs which generates an appropriate inter-scale interconnection of the extracted lines.

In the following section we describe the overall algorithm. Afterwards, we explain the newly developed details and we finish with some examples. We give analytical examples to validate that the feature vector is computed correctly and a medical example to show the potential of a classification algorithm based on the generated values.

1 Overall Algorithm

Several steps have to be performed in order to detect the line structures in the data and to generate the feature vector. First, the local maxima of the wavelet transform have to be computed for every scale. These maxima represent the isolated points shown in Figure 1(a). The coefficients for these maxima are used later on for the calculation of the feature vector. Second, the local maxima of each scale are interconnected into lines using graph theoretical aspects (Figure 1(b)). The interconnection ensures stability of the segmentation process. As lines on different scales may correspond to the same feature in the image, the interdepencies of the lines across different scales have to be established. In Figure 1(c) all lines, that have a correspondence at the second scale are drawn black, and the others are drawn grey. Finally, the feature vector for a line on the first scale can be computed using the average wavelet coefficients of this line at the other scales. Each processing step will be described in more detail in the next sections.

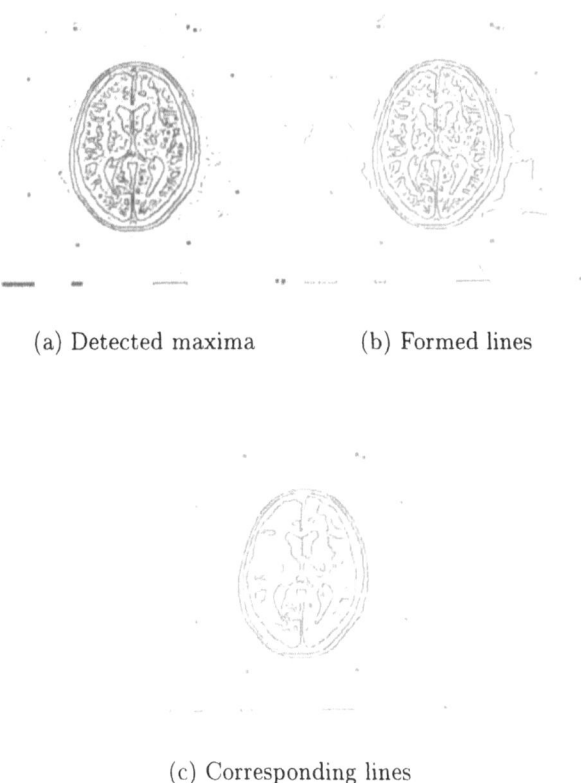

(a) Detected maxima   (b) Formed lines

(c) Corresponding lines

Fig. 1. Stages of the segmentation process

2 Construction of Local Maxima

In this section we describe how to find some pixels as atomic structures, which correspond to the local maxima of a continuous wavelet transformation. The wavelet transformation uses two wavelets which are obtained by differentiating the Gauss function in both $x-$ and $y-$ directions. This actually corresponds to using the Canny-Operator for different scale factors.

The multidimensional information which constitutes the feature vector is based on analytical aspects of the underlying function, such as Hölder continuity, gradient absolute value, etc. The wavelet theory provides a framework for analyzing these quantities. The one dimensional wavelet transformation of a function $g \in \mathbf{L}^2(\mathbf{R})$ for a wavelet ψ is defined using the convolution operator

$$(L_\psi g)(a, b) = \frac{1}{a} \int_{-\infty}^{+\infty} g(t)\psi\left(\frac{t - b}{a}\right) dt,$$ (1)

where b is a translation parameter and a a dilatation parameter. For two dimensions the continuous wavelet transformation may be defined using two wavelets ψ^1 and ψ^2, which are in our case the partial derivatives of the Gaussian function:

$$L_{\psi^1} f(a, x, y) = (f * \psi_a^1)(x, y)$$ (2)

$$L_{\psi^2} f(a, x, y) = (f * \psi_a^2)(x, y)$$ (3)

Using the partial derivatives of the Gaussian function has the two advantages, that it is infinite continuously differentiable and that it has only one vanishing moment. This guarantees the smoothness of the filtered function on the one hand. On the other hand only singularities with a Hölder exponent smaller than one will be detected, which is a standard requirement for image processing. Furthermore, we are only interested in finding discontinuities of the underlying function. The Hölder exponent is defined as follows: A function g is Hölder continuous α over an open finite interval Ω if and only if

$$\exists A : \forall x, x_0 \in \Omega : |g(x) - g(x_0)| \le A|x - x_0|^\alpha .$$ (4)

The Hölder exponent describes the continuity qualities of the function. It provides a more differentiated way to express continuity than the standard C^n continuity expression. As edges are singularities of the image, they can be characterized by this exponent. The following theorem describes the relation of the Hölder exponent and the wavelet coefficients.

The function $g \in \mathbf{L}^2(\mathbf{R}^2)$ is uniformly Hölder continuous α on a finite open set Ω if and only if

$$\exists A : \forall(b \in \Omega) : \forall a : |L_\psi g(a, b)| \le Aa^\alpha$$ (5)

holds, where $L_\psi(a, b)$ means the wavelet transformation for the wavelet ψ for translation b and dilatation a (see [7, 8, 9]).

In order to use the results of this theorem one first has to determine the domain, where the Hölder exponent has to be evaluated. Furthermore, one has to find the wavelet coefficients which make (5) as tight as possible.

We are interested in regions around the edges or lines, which are given by local wavelet transform maxima. Because the Gaussian filter tends to disturb and to blur the structures in the data, when the variance becomes large, we will consider such structures which can already be seen for a dilatation parameter $a = 1$. Since one edge has different representations in terms of maxima lines for different dilatation parameters, we consider the problem of building lines from the wavelet transform local maxima in the next section.

3 Construction of the lines

In order to construct these lines from the isolated pixels graph theoretical methods are used. An important aspect of the algorithm presented is, that the criteria for operating on the graph to obtain structures which are called lines are motivated by the Gestalt psychology.

Initially, a fully meshed graph is constructed. This graph is divided into a minimal spanning tree. In this work a simpler method has been developed, that performs the construction of the tree from the originally unconnected nodes in one step. This is done using an advancing front technique. Three kinds of nodes are distinguished: nodes that are already *fixed*, nodes that still have not been processed or are *untouched* and nodes that have been processed or *touched*, but whose father can still change. The basic idea of the algorithm is to change one node from *touched* to *fixed* at every iteration. For each one of these nodes the closest k neighbors that are not fixed are analyzed. If their status is *untouched*, they change to *touched* and their father will become *fixed*. If their status is already *touched*, then the length of the root node is compared with the distance of the analyzed *fixed* node. If this distance is shorter, the root of the *touched* node will be changed to the analyzed *fixed* node. This is procedure illustrated in Figure 2.

This tree is separated into several subtrees using the statistical Z-test for the length of each edge and the difference of the wavelet coefficients of the bordering maxima. The processing of the edges works recursively from the root to the trees. If any of the specified values of the edge differs significantly from the mean of the Z-Test of the other edges that lie within a specified hops distance, the tree is truncated into two clusters. The accumulation of the edge values is again done using a recursive mechanism, that searches forward in the direction of the leaves and backwards in the direction of the root.

The generated clusters are separated into lines afterwards. This is done using a gradient criterion and a length criterion for the lines. Every junction is considered. All lines that are closer to the gradient direction than a given angle are eliminated. If the junction still exists after the elimination, then all lines are separated from the junction that does not belong to the longest possible line, that passes through it. These techniques are described in Zahn [12]. The result of this phase is a set of lines for every scale.

4 Inter-scale Interconnection

For the evaluation of equation (5) all the corresponding wavelet coefficients for each scaling factor are needed. This means that an interconnection of the lines

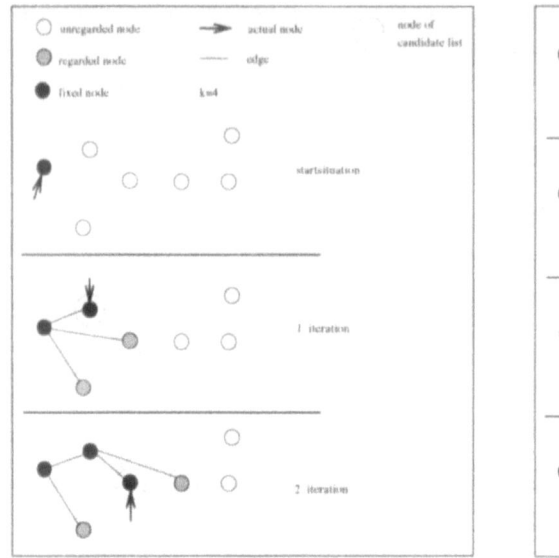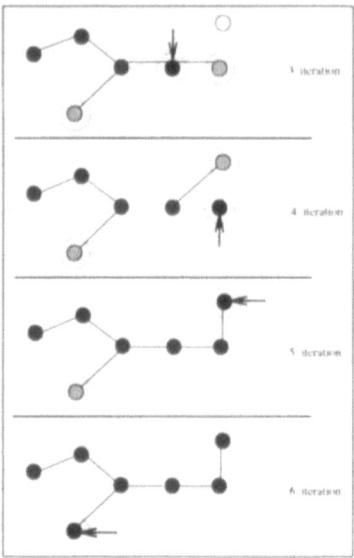

Fig. 2. Construction of the minimal spanning tree

which belong to the same edge has to be established for different dilatation parameters.

Experimental results have shown, that a simple interconnection between lines is not sufficient, because of the perturbations introduced by the convolution with the Gaussian filter. Depending on the dilatation parameters, lines may be grouped forming completely different structures at different scales. This problem has been solved by means of a carefull regrouping of the lines. The interconnection of the lines between neighboring scales is interleaved with a splitting of the lines at each scale to avoid meaningless interconnections.

For this purpose an initial point to point interconnection, based on the Euclidean distance, is computed. Every node at the scale $n + 1$ is connected to its nearest node at the scale n. If two nodes at the scale $n + 1$ are connected to the same node at the scale n, only the shortest link will be kept. This corresponds to a kind of discrete Hausdorff distance.

In a second step a region growing is performed on the lines. The partner lines of the connected points are considered as a homogeneity criterion. The best pairs of already fused segments are fused again. The process stops, when all of them have reached a minimum size and if no further fusion operation can be carried out without corrupting a minimum homogeneity. The resulting points are then fused to segments, that are parts of the lines. New lines that are interconnected across different dilatation parameters are grouped from these segments.

5 Computation of the feature vector

When the interconnections have been completed, five characteristic values for each discovered edge can be computed:

– the Euclidean length of the edge,
– the Hölder exponent (α),
– the variance where the edge has its last representation,
– the strength, whixh is the factor A in the equation,
– and an unsharpness value (σ), that may be extracted if further assumptions to equation (5) are made (see [8]).

The three characteristic quantities unsharpness, strength and the Hölder exponent are orthogonal quantities. If we assume that the image being considered is the result of a convolution of an unknown image with a convolution kernel of the variance σ and if we insert this assumption into equation (5) we get:

$$ln(L_\psi g(a, \mathbf{b})) \leq ln(A) + ln(a) + \frac{1}{2}(\alpha - 1)ln(a^2 + \sigma^2) \qquad (6)$$

This equation contains all the relevant entities which can be derived, if enough wavelet values for different scales are known. Most of the features are computed by means of a least squares approximation to make the resulting inequality as tight as possible. For stability reasons, the mean values and not the maxima of the wavelet coefficients for each dilatation parameter are used. This approach is justified by the fact that there should not be too rapid changes in wavelet coefficients along lines due to the pruning criterion in the graph processing section.

Experiments with a Gauss-Newton approximation have been made, but the approximated function turned out to be too unstable. The gradient descent method has been used instead as suggested in [8].

6 Results

In order to illustrate the correctness and the potential of the presented algorithm, two examples are discussed in this section. First we will evaluate the correctness of the discovered values by means of test images. Figure 3(a) is constructed using grey values of 50, 75 and 100. The horizontal line has an Hölder exponent of 0, a strength of 39,89 and an unsharpness of 0. The algorithm discovers the lines shown in Figure 3(b) and computes the following feature values for five evaluated scales: Hölder exponent -0.02, strength 36.71 and unsharpness 0. For seven evaluated scales the results are in even better correspondence: Hölder exponent 0.04, strength 37.4 and unsharpness 0.

As an example for segmentation we have chosen a slice of a MRI-head. Our aim was to segment the skin surface, the surface of the brain and of the ventriculars. We show the limits for the features and the segmentation results for two different MRI slices of two different heads. The skin surfaces in Figures of the color plate have been extracted with the following constraints to the feature

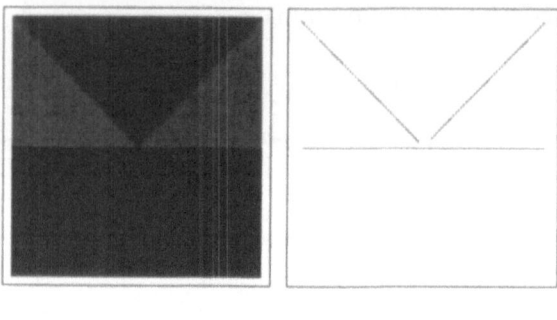

(a) Original image for
line detection

(b) Segmented line

Fig. 3. Analytical test data for image segmentation

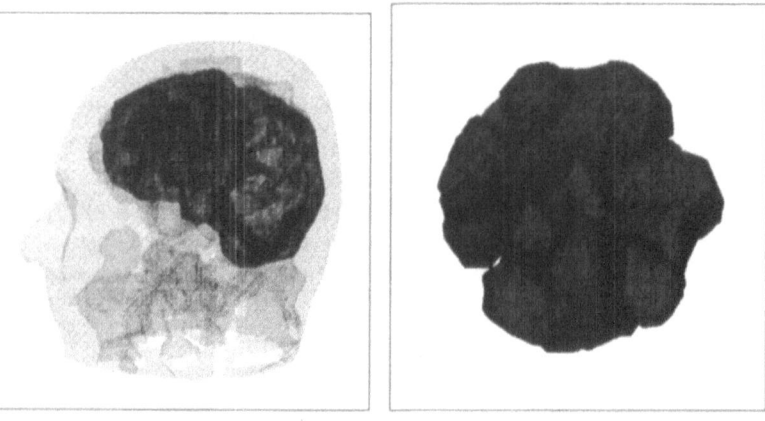

(a) Semi transparent rendering
of head and brain surface

(b) The ventricular system seen
from below

Fig. 4. Segmentation of a MRI volume

vector: length > 0.3, scale > 4, strength > 30. The brain surfaces are found with
length > 0.5, $2 <$ scale < 3 and the ventriculars have been extracted using: $0.01
<$ length < 0.6, scale > 3, $10 <$ strength < 36, $-1<$ Hölder exponent < 0, $0.5 <$
unsharpness < 2.5

Based on our encouraging 2D segmentation results we started extending these
techniques to three dimensional applications. Here again, the wavelet transform
maxima are used as a starting point for the segmentation process. In three
dimensions the graphs can not be used directly, since the result of this technique
are lines and not surfaces. In this case, the vertices are interconnected using a
Delaunay tetrahedrization. The Delaunay tetrahedrization guarantees, that no

other vertices are positioned within the circumsphere of a tetrahedron, than the vertices belonging to the tetrahedron.

The extracted surfaces are surfaces of groups of tetrahedra. The final grouping of lines in the two dimensional case was done using a kind of region growing as described in section 4. In the three dimensional case the tetrahedra are also grouped using a region growing technique, where the radius of the circumsphere of a tetrahedron serves as a homogeneity criterion. This is a variation of the α-shape concept presented by Edelsbrunner [4]. In this case tetrahedra formations are build of tetrahedra, that have a circumsphere radius below a certain limit.

As a feature vector, the surface to volume ratio, the volume size and the average circumsphere radius is computed. The segmentation result of a MRI head with the brain surface is shown in Figure 4.

This presented technique may also be used for labeling voxels for direct volume rendering or for the integration of the extracted surfaces with direct volume rendering. The brain surface contained in an MRI, can not be visualized with an iso-surface since this would also produce surface elements in other parts of the head. The segmentation procedure, however, offers the potential to separate the brain. Transfer-functions for direct volume rendering can be based on the values of the feature vector instead of the values of the dataset itself since they are a much better characterization of tissue type than the measured absorption values.

7   Conclusion

We have shown that the combination of wavelet and graph theory results in a stable segmentation procedure. The results of this process can be used for visualizing medical data sets with quite more meaningful images, than the application of standard mapping methods as transfer functions or isosurfaces derived from the original scalar values. In the future we will extend our first attempts to 3D segmentation by combining α-shapes and region growing on the extracted edge information. Furthermore, mechanisms have to be found to determine ranges of characteristic values for interesting medical features.

References

1. K. Berkner, C. J. G. Evertsz, and W. Berghorn. Eine lokale charakterisierung von kanten basierend auf skalierungseigenschaften von wavelet-transfomationen. In Carl J.G. Evertsz, editor, *Conference Guide Visualisierung- Dynamik und Komplexität*, September 1995.

2. A. Bijaoui. Partial reconstruction and astronomical image inventory by the wavelet transform. In *Proceedings ICIAP 1993*, 1993.

3. J. Canny. A computational approach to edge detection. *IEEE Trans. Patt. Anal. Machine Intell.*, PAMI-8:679–698, 1986.

4. H. Edelsbrunner and E. P. Mücke. Three dimensional alpha shapes. *ACM Transactions on Graphics*, 13(1):43–72, January 1994.

5. J.A.F Leite and E. R. Hacock. Statistically combining and refining multi-channel information. In *Proceedings ICIAP 1993*, 1993.

6. M. Levoy. Display of surface from volume data. *IEEE Computer Graphics and Applications*, 8(3):29–37, 1990.

7. S. Mallat and W.L. Hwang. Singularity detection and processing with wavelets. Technical report, Courant Institute of Mathematical Sciences New York University, march 1991.

8. S. Mallat and S. Zhong. Characterization of signals from multiscale edges. *IEEE Transactions on Pattern Analysis and Machine Intelligence*, 14(7):710–732, Juli 1992.

9. S. Mallat and S. Zhong. Wavelet transform maxima and multiscale edges. In *Wavelets and their Applications*. Boston: Jones and Bartlett Publication, 1992.

10. R. Westermann. Compression domain rendering of time-resolved volume data. In *IEEE Visualization 1995 Conference Proceedings*, pages 51–58, 1995.

11. A. Witkin. Scale space filtering. In *Proc. Int. Joint Conf. Artificial Intell.*, 1983.

12. C. T. Zahn. Graph-theoretical methods for detecting and describing gestalt cluster. *IEEE Transations on Computers*, C-20(1):68–86, 1971.

Editors' Note: see Appendix, p. 182 for colored figures of this paper

Interactive Segmentation and Analysis of Fetal Ultrasound Images

Kalpathi R.Subramanian   *Dina M. Lawrence*   *M. Taghi Mostafavi*

Department of Computer Science
The University of North Carolina at Charlotte
Charlotte, NC 28223
{krs,dmlawren,taghi}@mail.cs.uncc.edu

Abstract The ability of ultrasound scanners to image anatomical structures in real time have led to their use in two important applications of medicine, (1) for monitoring the unborn baby (fetal ultrasound), and, (2) coronary treatment of blockages in blood vessels (intravascular ultrasound). The generated images (in the form of a continuous video) are typically noisy and contain numerous artifacts, making it difficult to isolate and measure features of interest. We explore the use of two algorithms, region growing and a variant of split/merge algorithm for segmenting sequences of fetal ultrasound images. We describe an interactive system that can rapidly process and segment an arbitrary number of features. The system exploits frame to frame coherence for accelerating the segmentation process, while at the same time combining the strengths of these algorithms and some post-processing for accurate and robust detection of features.

1   Introduction

The use of medical imaging scanners for routine clinical diagnosis and treatment planning has gained wide acceptance in medicine in recent years, primarily because the technology is non-invasive (CT, MRI) or minimally invasive (intravascular ultrasound). The most common imaging modalities have been CT (Computer Tomography), MRI (Magnetic Resonance Imaging) and Ultrasound. Ultrasound scanners have a number of advantages over CT and MRI. Perhaps the most important among them are their ability to visualize anatomy easily and at interactive speeds. Ultrasound has found wide application in fetal diagnosis and coronary treatment [6]. Today, almost all pregnant mothers undergo at least one ultrasound scan, during the gestation period, for preventive reasons or early detection of problems relating to the health of the baby or the mother. Coronary (or intravascular) ultrasound is targeted at detection and treatment of atherosclerosis, the buildup of cholesterol induced blockages in the coronary vessels.

Computer assisted processing, analysis and archival of ultrasound images presents new possibilities for improved diagnosis and treatment planning. Critical segments of an ultrasound scan could be processed and archived for review at a later stage, or for training or instructional purposes. In this age of information technology, such data could be made available to physicians at remote sites (via Internet, for instance). Archiving such data using computer technology (as opposed to analog media such as video tape or film) would also make it a more efficient process, since only important segments of a scan need to be archived, and retrieval is considerably faster.

Processing ultrasound images for feature identification and analysis poses difficulties (in contrast to CT or MRI) due to the poor quality of the images[3][4]. Ultrasound images are noisy, possess poor contrast, suffer from variations in illumination and from self shadow problems that result in masking features of interest. The constant motion of the baby is another distinguishing problem faced by computer based processing, unlike CT and MRI scans (with the exception of the beating heart). Motion induced artifacts need to be compensated for before application of any image processing operations to analyze the images. Conventional image processing operators often work very poorly on ultrasound images. Thus, it is important to investigate robust techniques that can reliably enhance and process ultrasound data, for analysis and quantification of anatomical features of interest.

Earlier work on ultrasound image processing and analysis has focused on 3D reconstruction [8, 10, 12, 11]. Unlike CT or MRI, where 3D data can easily be acquired from tomography machines (by stacking a sequence of parallel 2D images), ultrasound scanners produce a video signal that needs to be discretized to obtain 2D images; however, these images are arbitrarily oriented and special 3D sensors will need to be used to obtain spatial information to orient and locate each 2D image [8]. The problems associated with this process have limited the use of such methods to medical research centers or hospitals.

In this article, we describe the development of a new 2D interactive system targeted at processing and segmenting sequences of fetal ultrasound images. The project, in collaboration with Carolinas Medical Center (CMC) [1], is aimed at providing technicians a simple tool that would allow rapid identification and quantification of features of interest over selected segments of an ultrasound scan. The system exploits two ideas for rapidly processing long sequences of 2D ultrasound images, (1) frame to frame coherence, and (2) allow the user to specify regions of interest over which the processing algorithms operate. In this work, we have investigated region growing and a variant of the split/merge algorithms, individually and in combination to segment ultrasound images. We have found out that using the two algorithms in combination with some postprocessing produces more accurate segmentation of the ultrasound images. Preliminary tests of the system indicate that it is possible to segment sequences of up to 50 images (with 2-5 regions of interest) in about 5-6 minutes on standard Unix workstations.

2 Background

2.1 Image Acquisition

Unlike CT and MRI, ultrasound scans are made using high frequency sound waves. In OB/GYN applications, an external probe is used as the source. Reflections of the emitted waves are measured by the sensors in the probe used to perform the scan. Materials with differing impedances cause reflections of varying intensities, which, when combined with their time of arrival at the sensor, helps in generating a 2D image. The strength of the reflection determines the intensity of the image pixels, while their arrival time determines their spatial location in the image. The received signal undergoes internal processing and is then converted into a video format (NTSC, RGB, component, etc) for display on a standard monitor or for recording. The raw images are usually grayscale images, although some scanners provide pseudo coloring for highlighting and labeling various features in the data.

2.2 Processing and Segmentation

For a variety of reasons, ultrasound images contain numerous artifacts [5, 1] that poses significant challenges to computer processing. Poor resolution (axial and lateral), reverberation (multiple reflections) and shadowing are some causes of artifacts in ultrasound

---

[1]CMC is a statewide hospital authority in Charlotte, NC and includes a dedicated research center.

images. The motion of the baby further complicates the processing algorithms, introducing considerable blurring as well as loss of coherence from frame to frame. With the amount of noise present in ultrasound images, it is essential to perform some amount of smoothing prior to segmentation and feature detection. Both gaussian and median filters can be used to compensate for noise [11]; median filter has been found to be effective in removing speckle noise, which is common in these images. The kernel size to be used is largely based on trial and error, as too large a kernel overly blurs the images with corresponding loss of contrast. Some level of smoothing will need to be applied, else the algorithms that are targeted at feature identification will be highly sensitive to noise. Compensation for motion induced artifacts will be addressed in a companion paper. It is an important issue that needs to be dealt with properly, especially when attempting to segment the heart.

Once the image has been compensated for noise, the next step is to identify objects in the image. The process begins by segmenting the image into disjoint regions. Segmentation is a very difficult problem, and, to date, automating this process has been restricted to the simplest of all images or to organ and disease specific images. Thus, some form of human intervention is needed to assure a high level of confidence in this procedure. An extensive review of segmentation methods can be found in [2].

Two popular techniques to perform segmentation is through 'region growing' and the 'split/merge' algorithms [4]. In the region growing algorithm, a seed pixel is chosen by the user and its intensity is compared to those within a small neighborhood. A similarity measure (intensity tolerance, for instance) decides if any of the neighbors belong to the region of the seed pixel. Accepted neighbors recursively test their neighbors in a similar manner, resulting in a gradual growth of the region towards its boundary. The sensitivity of the algorithm directly depends on the sharpness of the region boundaries. Any weak or missing edge points on the boundary can cause the region to 'bleed' into neighboring regions.

In the split/merge algorithm, a section of the image containing the region of interest is selected by the user. A homogeneity condition is applied to the region's pixel intensities to check if they are sufficiently similar (for instance, the deviation from the mean pixel intensity of the region could be tested to be within a threshold). If this test fails, the region is partitioned into some number of equal sized partitions (a quadtree partitioning is common) and the homogeneity measure is applied to the partitioned regions. This process continues until all regions are classified to be homogeneous. Since the partitioning is performed in a somewhat independent fashion, it is very likely that adjacent regions of two independently partitioned regions could possibly be merged into a single region. Thus, the splitting procedure is followed by a merging step, resulting in merging adjacent regions (which is also a recursive procedure, allowing merged regions to be merged again, if necessary).

The region growing algorithm is a topologically sensitive algorithm and highly sensitive to missing points in the boundary. In addition, it is incapable of detecting disconnected regions belonging to the same object. The split/merge algorithm is a more global approach to determining the regions of an object and uses statistical measures of a region to determine region membership. Such complementary properties of these two algorithms might be used in combination to produce a more robust and accurate segmentation method. At the same time, due to variations in illumination and other artifacts in ultrasound images, both algorithms can leave holes within features, introducing errors in estimating feature size. Morphological operators [4, 3] can be used to compensate for this. The two basic operators, *erosion* and *dilation* can either erode or grow the boundary of a region by an amount determined by the structuring element used. Dilation can be used to fill small holes in images, while erosion can be used to reverse the effects of dilation (for maintaining the size of the region, for instance). Finally, the split/merge algorithm, because of its global nature, can classify disconnected clusters of pixels as belonging to the same object. Where applicable, this can be compensated by the use of a connected component labeling procedure (to be described in Secion 3.2).

Segmentation is typically followed by computing descriptors for the identified objects.

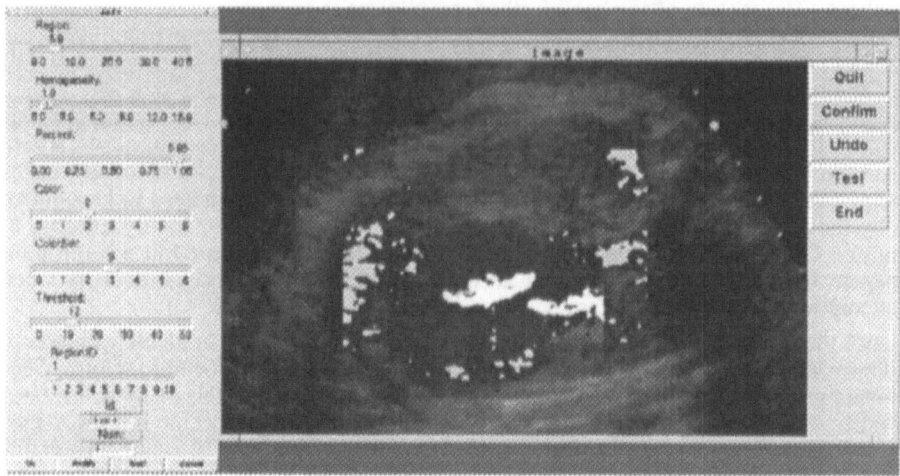

Fig. 1: 2D Segmentation System Interface

Descriptors could be based on the boundaries of the objects, region based, topological or textural. For ultrasound images, quantification of the size, location and orientation of objects in the image are useful descriptors.

3 2D Segmentation System Description

The current implementation of the system is on Silicon Graphics workstations; A snapshot of the user interface is illustrated in Fig. 1, when using the overlay algorithm, which allows application of both region growing and split/merge in sequence. The user interface is built using the Tcl/Tk toolkit [9], which is publicly available for a variety of architectures, including PCs. All of the processing modules are implemented as Tcl commands and are called from the interface.

3.1 The User View

A user session begins by reading in a sequence of acquired ultrasound images (individual images may also be read separately). The sequence can be smoothed using either a Gaussian or median filter. The user needs to specify the width and deviation of the Gaussian kernel or just the width for the median filter. Once the smoothing is completed, then the sequence can be segmented using (1) Region Growing (2) Split/Merge, or (3) Overlay, where the two algorithms are applied successively, in any order. In each case a rectangular region of interest is first selected to restrict the area of the image over which the segmentation algorithms operate. After a segmentation algorithm is chosen, the region is segmented. Any number of regions can be segmented in this manner. Once the segmentation is completed for the current image, the user can move on to the next image in the sequence, at which point the parameters used to segment the current image are automatically applied to the new image. If this is satisfactory, the user continues onto the following image. If the segmentation is unsatisfactory, the user can enter a 'modify' state, where any of the regions can be segmented again by modification of the parameters controlling the segmentation of the region. The system maintains a list of regions, with associated lists of points making up the region; these are efficiently manipulated to reflect the changes in segmentation of the region. Since the segmented regions typically contain holes due to variations in illumination, we have also implemented morphological operators (erosion and dilation) as a means to filling them. In particular, dilation has been found to be useful as a postprocess to fill holes left by region growing and split/merge. At any point, a segmented image may be written out to secondary storage for later review and playback. Once the segmentation of a image sequence terminates, the system outputs statistics that reflect the characteristics of the region. In the current implementation, once the processing is completed for each image, the total number of pixels making up each region, the mean intensity and standard deviation is output.

3.2 Segmentation Algorithms

In the region growing algorithm, the user interactively selects a seed pixel from any point within the region of interest and its intensity is determined. A tolerance value (in intensity units) is specified through the interface (threshold parameter in Fig. 1). The four orthogonal neighbors of the seed pixel are next evaluated for membership; a pixel is added to the region if its intensity is within the specified intensity tolerance, i.e.

$$I_{seed} - I_{tol} \leq I(x, y) \leq I_{seed} + I_{tol}$$

where I_{seed}, I_{tol} and $I(x, y)$ are the intensities of the seed pixel, specified tolerance and the intensity of the pixel being tested for region membership. A pixel accepted as part of the region will recursively have its neighbors tested. For efficiency, our implementation uses a queue to determine the accepted pixels.

The split/merge algorithm used in our implementation is a modified version of the standard algorithm. The modification makes the algorithm sensitive to the properties of the region of interest. The algorithm begins by determining the mean intensity, μ_r, and standard deviation, σ_r of a small rectangular neighborhood (user defined, 11×11 in our implementation) region r surrounding a chosen point within this region. Next the algorithm performs a quadtree partitioning of the region, creating 4 equal sized sub-regions. Each of these sub-regions, s_i is evaluated for homogeneity. Homogeneity is determined by first determining the mean and standard deviation, μ_s and σ_s of the sub-region. A subregion is homogeneous if a given percentage p_s of pixels in the subregion have intensity values which fall within n_s number of standard deviations of the subregion's mean. In our implementation, all of these parameters are controlled through the interface (Fig. 1). Typically, $p_s = 0.95$ and $n_s = 1$. If a subregion is determined to be homogeneous, then its statistics are compared to the properties of the region of interest, μ_r and σ_r. If μ_r is sufficiently close to μ_s, all pixels of s_i become part of the segmented region. For acceptance, $|\mu_s - \mu_r|$ must fall within n_r standard deviations of the mean of the region of interest. Thus, there are two tests a subregion must pass before acceptance:

1. $|I(x, y) - \mu_s| \leq n_s \sigma_s$, must be satisfied by at least p_s percent of the pixels of sub-region s_i.

2. $|\mu_s - \mu_r| \leq n_r \sigma_r$.

If both of these conditions are satisfied, then the sub-region s_i is accepted and the region is colored. As our implementation currently keeps an explicit representation of the points making up the segmented region, explicit region merging is unnecessary.

Finally, our segmentation system is also capable of applying both algorithms in sequence on a particular region (which we call an overlay), in an attempt to take advantage of their combined strengths. Our experimental results have indicated some success to using both algorithms in sequence, most notably to fill in holes that are generated by the region growing algorithm. A connected component labeling procedure to discard unrelated parts of the region has been found to be useful with the overlay algorithm. This procedure performs region growing with the same seed pixel, with the exception that

Table 1: Test Results/Runtime Statistics.

Dataset	Image Size	#frames	#Regions	Time (minutes)			
					Segmentation		
				Smooth.	Reg. Grow (#modified)	Spl. Mrg. (#modified)	Overlay (#modified)
1. Face-1	450×590	30	2	5	3(2)	3(3)	5(3)
2. Heart-1	640×400	30	5	5	6(5)	5(8)	7(8)
3. Face-2	230×160	50	2	1	5(6)	6(11)	5(3)
4. Heart-2	260×260	50	2	1	6(9)	8(8)	9(8)
5. Face-3	210×160	49	2	2	5(5)	5(8)	4(2)
6. Heart-2	325×235	50	3	2	8(14)	7(9)	10(9)

only pixels within the region that are connected to one another (4-connected neighborhood, starting with the seed pixel) are reported as being part of the segmented region. Any remaining holes in the region can be eliminated by applying a dilation operation on the region. Dilation is currently performed using a 3×3 structuring element of unit height.

4 Experimental Results

We have digitized (from video) six sequences of ultrasound images, obtained from three different ultrasound scans. Datasets 1 and 2 are from a scan of an 8 month old fetus. An Accom video digitizer was used to obtain two segments (Face-1 and Heart-1) of the scan corresponding to the baby's face and heart. The second scan was of a baby that was 17 weeks old. An Abekas Diskus digitizing recorder was used to obtain 3 sequences of images (Face-2, Heart-2 and Face-3), corresponding to the baby's face, heart and a second view of the face. Finally, the third scan is of a 5 month old baby, from which 1 segment was digitized, again using the Abekas Diskus and corresponds to the baby's heart.

Table 1 gives details of the test data and processing times for our experiments. All of the processing was performed on a SGI Indigo-2 Impact workstation with an R10K processor. All datasets were smoothed using a 7×7 median filter except for Heart-2 which used a 5×5 sized kernel. Segmentation times are comparable for all three algorithms. In the overlay algorithm, the region growing algorithm is followed by split/merge. The number of images that needed modification (or to be segmented again) is also shown and ranges anywhere from 4 to 18% of the total number of images in the sequence.

It can be seen that frame to frame coherence helps accelerate the segmentation of the image sequences. For longer sequences, the average time to segment each image should further decrease. Note that in the case of Face-2 and Face-3, the overlay algorithm results in modification of fewer images than either of the two algorithms applied individually.

Table 2 reports on the sizes of the segmented regions. A sample image from 3 of the datasets was used to illustrate the behavior of the various algorithms. Columns 3 and 4 show the region sizes due the region grow and split/merge applied individually. In general, the size reported by split/merge is larger, which is mostly due to unrelated regions, although it does fill in some of the holes left by the region growing algorithm. The last two columns show the region sizes resulting from the overlay algorithm followed by the connected component labeling procedure (column 5), which is followed by dilation (column 6). This sequence of segmentation procedures allows us to be conservative in the region growing and split/merge procedures; while this tends to underestimate the

Table 2: Region Sizes.

Dataset	Region No.	Region Size (pixels)			
		Reg.Gr.	Spl.Mrg.	Conn. Comp.	Dil.
1. Face-1	1	3286	3371	3571	4102
	2	2695	2378	2761	3240
2. Heart-1	1	666	764	865	1059
	2	375	411	397	457
	3	18873	19043	19161	20754
	4	1623	1598	1806	1897
3. Face-2	1	334	364	433	633
	2	571	741	748	286

feature size, it is necessary to prevent bleeding; dilation helps fill most of the holes that remain in the regions as well as grow the boundary in a controlled fashion. The use of several procedures does increase the user interaction time, but it permits a more accurate segmentation of the region.

In the top row of Fig. 2 (color plate in Appendix) are shown examples from Face-1. The left image is the original (which has been cropped), the middle image shows the eye sockets segmented by region growing, and the right image also includes dilation. The dilation helps close most of the holes in the segmented region. The second and third rows illustrate the effect of the various algorithms on an example image from the sequence of Heart-1. The left image is the original, middle image shows 4 regions segmented using region growing and right image uses the split/merge algorithm. The white regions (3 of them) are part of the wall that separate the upper and lower chambers of the heart. In the third row, the left image shows the segmentation using the overlay algorithm (region growing followed by split/merge), the middle image shows the same with the disconnected regions eliminated by application of the connected component algorithm. The right image is the result of dilating the segmented regions in the middle image using a 3×3 structuring element (box of unit height). This helps fill the holes in the regions and provides a better estimate of the feature size. In our experimentation, we find that the region growing algorithm by itself tends to be conservative in estimating the size of the feature; increasing the tolerance to include more of the region typically results in including points outside the region, causing it to bleed. The split/merge method helps fill in some of the holes (yellow region in the overlay example) and add more pixels to the outside of the region. But it does generate a number of pixel clusters that are unrelated to the feature, due to the global nature of the algorithm. The connected component procedure eliminates all of these and the dilation is very effective in filling the holes that are still left behind (the white region in the middle and right images of row 3 indicate pixels that are common to both region grow and split/merge). Finally, the left image in row 4 illustrates the segmentation (using region growing only) of the heart into its 4 chambers with the walls in white[2]. Since neither the wall nor the valve is well defined in the original (left image, row 2), it is very difficult to get a clean segmentation of all of these features. The top and bottom images of row 4 are examples from Face-2 and Heart-2 respectively. Here again, the two eye sockets are segmented in Face-2 and two regions corresponding to the heart of Heart-2.

---

[2]A Quicktime animation of a sequence of the segmented heart may be found in http://www.cs.uncc.edu/~krs/publ.html

122

5 Concluding Remarks/Future Work

In this article, we have explored the use of two algorithms, region growing and a variation of split/merge for segmentation of fetal ultrasound images. Because of the variety of artifacts present in ultrasound images as well as their poor dynamic range, we have found that it is almost impossible to apply automatic segmentation methods with a high degree of confidence. As a result, we have embarked on an interactive approach, with mechanisms to speed up the segmentation via frame to frame coherence. We are able to segment sequences of 30-50 images within five to six minutes, depending on the degree of coherence between consecutive images. Considering the human interaction involved, the ability to segment each sequence in such a short time is quite promising. In the cases where the boundaries of the regions are fairly well defined, we have been able to use the region growing or split/merge algorithm to successfully segment the images. Our experimentation with both region growing and the split/merge algorithms leads us to conclude that their combined application to ultrasound image sequences, followed by some post-processing to overcome the weaknesses of these algorithms results in a more accurate and robust detection of features. The biggest weakness of the system in its current form is the lack of effective measures that can evaluate the accuracy of the segmentation, and hence, the results tend to be somewhat subjective. One idea to address this is to use ultrasound technicians to interactively mark the features of interest and evaluate the results.

This work highlights the difficulties involved in reliable segmentation of ultrasound images. It also calls for techniques that are more tolerant of the noise and artifacts present in ultrasound images. Algorithms such as region growing are highly sensitive to the local neighborhood, requiring sharp discontinuities (boundary) for region isolation. A better characterization or quantification of the boundary would help in improving the noise tolerance of these algorithms.

A representation that provides a better characterization of the boundary is the *Binary Space Partitioning (BSP) tree* [7]. BSP trees represent functions by encoding all of the discontinuities within a function via partitioning hyperplanes. The representation is simply a binary tree, with internal nodes containing hyperplanes (representing discontinuities) and the leaf cells representing regions that are relatively homogeneous. These regions can be approximated by a continuous function, such as a constant (say, the mean of the pixel intensities) or a linear function.

BSP trees have been used to represent CT and MRI images [13]. They have been shown to be highly tolerant to noise and missing boundary points [14], and hence could find application to ultrasound images. The method used to convert the discrete image into a partitioning tree (refer to [13] or [14] for details) makes it possible to reliably identify the image discontinuities (or boundary), even in the presence of considerable noise or missing edge points. Thus, performing the representation conversion to a BSP tree and then applying algorithms such as region growing on the continuous representation should lead to a more robust segmentation algorithm.

6 Acknowledgements

Our thanks to Carolinas Medical Center, Charlotte, for providing us with clinical data that was used in the experiments. This research was supported, in part, by the National Science Foundation, and by a faculty research grant from UNC Charlotte.

References

[1] R.J. Bartrum and H.C. Crow. *Gray-Scale Ultrasound: A manual for Physicians and Technical Personnel.* W.B. Saunders Company, Harcourt Brace Jovanovich, Inc., 1977.

[2] J.C. Bezdek, L.O. Hall, and L.P. Clark. Review of mr segmentation images using pattern recognition. *Medical Physics*, 20(4):1033–1048, 1993.

[3] C. Busch and M. Eberle. Morphological operations for color-coded images. *Computer Graphics Forum*, 14(3), 1995.

[4] R.C. Gonzalez and R.E. Woods. *Digital Image Processing*. Addison Wesley, 1992.

[5] F.W. Kremkau. *Diagnostic Ultrasound*. W.B. Saunders Company, Harcourt Brace Jovanovich, Inc., 1989.

[6] J. Lengyel, D.P. Greenberg, and R. Popp. Time-dependent three-dimensional intravascular ultrasound. *ACM Computer Graphics Proceedings*, August 1995.

[7] B.F. Naylor. Interactive solid modeling using partitioning trees. In *Proceedings of Graphics Interface '92*, Vancouver, CA, May, 1992.

[8] T.R. Nelson and T. Elvins. Visualization of 3d ultrasound data. *IEEE Computer Graphics and Applications*, 13(6), November 1993.

[9] J. K. Ousterhout. *Tcl and the Tk Toolkit*. Addison Wesley, 1994.

[10] J.R. Roelandt, C. di Mario, N.G. Pandian, L. Wenguang, D. Keane, C.J. Slager, P.J. de Feyter, and P.W. Serruys. Three-dimensional reconstruction of intracoronary ultrasound images. rationale, approaches, problems, and directions. *Circulation*, 90(2), August 1994.

[11] G. Sakas, L. Schreyer, and M. Grimm. Preprocessing, segmenting and volume rendering 3d ultrasonic data. *IEEE Computer Graphics and Applications*, 15(4), July 1995.

[12] G. Sakas and S. Walter. Extracting surfaces from fuzzy 3d-ultrasound data. *ACM Computer Graphics Proceedings*, pages 465–474, August 1995.

[13] K.R. Subramanian and B.F. Naylor. Representing medical images with partitioning trees. In *Proceedings of Visualization '92*, Boston, MA, Oct. 19-23, 1992.

[14] K.R. Subramanian and B.F. Naylor. Converting discrete images to partitioning trees. *IEEE Transactions on Visualization and Computer Graphics*, 1996. Submitted.

Editors' Note: see Appendix, p. 183 for colored figure of this paper

Efficient Visualization of Large–Scale Data on Hierarchical Meshes

R. Neubauer[1], M. Ohlberger[2], M. Rumpf[1], and R. Schwörer[2]

[1]Institut für Angewandte Mathematik, Universität Bonn,
Wegelerstr. 6, 53115 Bonn
[2]Institut für Angewandte Mathematik, Universität Freiburg,
Hermann–Herder–Str. 10, 79104 Freiburg

Abstract. A multi-resolution approach is presented for data on a large class of hierarchical and nested grids. It is based on a procedural interface and a set of hierarchical and adaptive visualization methods. Such a method consists of a recursive traversal of mesh elements from the grid hierarchy combined with an adaptive stopping according to some error indicator which is closely related to the visual impression of data smoothness. During this traversal user data is only temporarily and locally addressed on single elements. No in advance mapping onto prescribed formats is necessary. The user only has to supply a set of element access routines as an interface to his specific data structures. As no extra storage is required, also large, economically stored computational grids can be handled on workstations with moderate local memory. Significant examples illustrate the applicability and efficiency on different types of meshes.

Introduction

Efficient numerical algorithms such as multi grid methods are nowadays capable to resolve complex structures in the simulation of physical processes. In a post processing step the user wants to explore the large amount of data with typically millions of unknowns interactively to improve the understanding of interesting features. Therefore efficient visualization tools are essential to extract the requested information from the enormous data base at a high frame rate.

The numerical methods are mostly based on a variety of domain discretizations such as structured or unstructured Finite Difference, Finite Element or Finite Volume grids, which are in general supplied with a hierarchical structure. These meshes may consist of a single or of mixed element types, e. g. simplicial, prismatic, rectangular or cuboidal, and they are frequently generated by different recursive, adaptive refinement strategies. Thereby non standard and application dependent data structures are often essential for an efficient implementation of the simulation algorithm.

The hierarchical type of these numerical data structures, first used for computing, is also well suited to improve the efficiency of a class of typical visualization methods. In the present paper two main aspects will be discussed.

First, we ask for a flexible integration of the above large class of hierarchical data structures from the applications into a post processing environment. The gap between the user's numerical data formats and the prescribed structures usually used by visualization tools is one of the fundamental outstanding problems in scientific visualization [7,15]. Most of the visualization software currently in use works on prescribed data formats [2,5,9,16]. User data has to be converted into such a format. But this is time and storage expensive especially in case of large nested grids, where very often closely related to the specific application an economical data storing is possible. It seems to be impossible to set up a fairly general and efficient data format covering all the above grid types. For non hierarchical meshes in [13] a different approach, which tries to avoid these difficulties, is proposed. A mesh is defined as a procedurally linked list of elements. There is no random access to a single element. Information about elements is only locally and temporarily provided by user supplied access procedures. This concept can be generalized to hierarchical grids. In section 1 we will introduce access routines to hierarchical elements supporting a recursive traversal of any nested grid hierarchy. Furthermore in section 2 a type of economical data structures to store nested grid data efficiently is discussed. We point out that such structures, which can exclusively been handled by a procedural approach, are well suited to store even very large grid geometries on a standard graphic workstation.

Second, especially the huge amount of data delivered by efficient numerical methods requires as well efficient visualization methods to support an interactive analysis of the physical characteristics modeled by the simulation. We will discuss mainly the extraction of isosurfaces as a typical graphical tool to inspect 3D data sets, although this methodology applies to other methods as well. The classical marching cube method, by Lorensen and Cline [11], efficiently renders local isosurfaces on hexahedral grids. But the underlying search algorithm for intersected elements is still an overall traversal of the set of elements. Mainly three different types of improvements have been investigated: an efficient presorting of elements [3,10,14], a seed point selection strategy combined with a spreading search for isosurfaces using adjacency information [6], and the recursive hierarchical search for intersected elements using precomputed and stored min/max values [17]. We sketch here how to combine a hierarchical search over the grid hierarchy with an adaptive stopping on elements where the visual improvement, one would obtain on finer grid levels, is below a user–prescribed error tolerance. Such an adaptive stopping on coarser grid levels has been considered for instance in the context of volume rendering for an adaptive splatting technique [8] and for an algorithm based on successively refined tetrahedral Delaunay meshes [1]. The advantage of our algorithm is, that it can be implemented on a fairly general class of hierarchical grids, which are in our case procedurally addressed, acts strictly local on elements without referring to their neighbourhood, and rules out discontinuities on the isosurface due to transitions between different grid levels. This concept can easily been generalized to other visualization methods, such as color shading or isoline drawing in 2D, or on volume slices in 3D. It applies to hierarchical meshes consisting of elements, which are tensor products of simplices in one, two or three dimensions and the corresponding function spaces generated by tensor products of linear functions on simplices. This especially includes the elements sketched in Fig. 1 with, for instance, linear, bilinear or trilinear functions defined on them. A detailed discussion of this concept can be found in [12]. Section 3 gives a brief overview on the basic ideas and in section 4 several examples underline the applicability of the presented concept. Let us finally

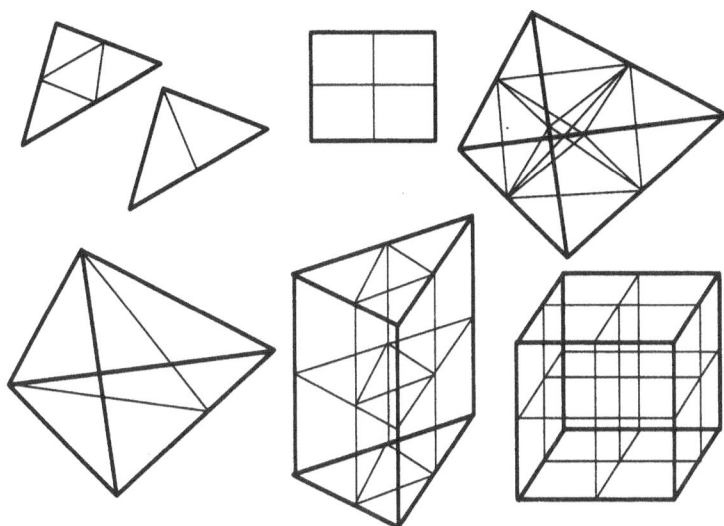

Fig. 1. The basic element types and their refinement (for triangles and tetrahedrons we depict two types of refinement).

clarify some notation concerning the grids, which will be used throughout the subsequent sections. A set of nested grids $\{\mathcal{M}^l\}_{0 \leq l \leq l_{\max}}$ is a family of meshes, which are recursively generated by refinement of certain elements of the preceding coarser mesh. More explicit, an element $E \in \mathcal{M}^l$ is refined according to some refinement rule and we thereby obtain a set of child elements $\mathcal{C}(E) \subset \mathcal{M}^{l+1}$ (cf. Fig. 1). By refining an element new vertices x are generated. Let us denote the set of vertices of E and $\mathcal{C}(E)$ by $\mathcal{N}(E)$, $\mathcal{N}(\mathcal{C}(E))$ respectively.

We finally remark that, although we mainly focus on the 3D case, most of the schematic figures deal with the 2D case, to simplify the presentation.

1 A procedural interface to hierarchical meshes

In [13] a visualization interface for arbitrary meshes with general data functions on them has been proposed. This interface tries to avoid restrictions on the element types. A mesh is defined as a procedurally linked list of non intersecting elements. The access to data is done by user supplied procedures addressing the user data structures and returning the required data temporarily in a prescribed element structure *ELEMENT*. It especially contains a reference to some element type, the coordinate vectors for the nodes, and function data on them. (Here we restrict ourselves to the basic concept. The true data structure is slightly more general, especially concerning the interface for function data). In general at the same time only one *ELEMENT* structure is present in storage. There is no random access to a single element. But this is in fact not necessary for most common and frequently used visualization methods (cf. section 3). Especially no permanent mapping of numerical data onto new data structures is required. The visualization tools directly work on the data structures the user is accustomed to from his numerical method. He only has to provide the access procedures and give a description of the element types. For the details on the procedural interface we refer to [13].

So far no hierarchical structure is taken into account. Now we enlarge this interface by access procedures which procedurally represent the tree structure underlying a nested grid. Two procedures *first_macro()*, *next_macro()* successively deliver information on the coarse grid elements in an *ELEMENT* data structure, overwriting previous element data. A call of *first_child()* generates and fills an additional *ELEMENT* structure with some first child data. Finally successive calls of *next_child()* traverses the other child elements of the same parent element and replaces previous child data (cf. Fig. 2). Thereby during a recursive traversal of the grid hierarchy, a list of at most n temporarily filled ELEMENT structures is present in memory at the same time, where n is the depth of the hierarchy. This especially implies that also in the user data structures the element information needs not to be stored completely on all grid levels, but it may be generated when needed, based on complete parent information and economically stored offset data (cf. section 2).

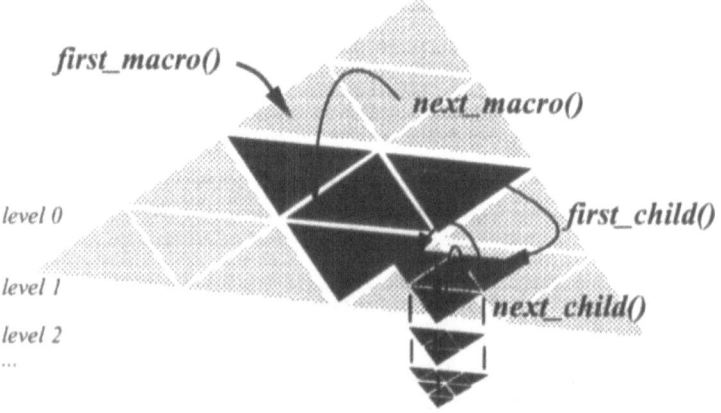

Fig. 2. A schematic sketch of the procedural access to hierarchical grids by the four routines *first_macro()*, *next_macro()*, *first_child()*, *next_child()*.

The above access procedures supply a visualization method with all necessary information to locally evaluate and graphically represent grid geometry and data. This is sufficient to run merely all visualization algorithms, e. g. isosurface rendering, isoline drawing and data dependent color shading. In section 3 we will see that additional error indicator values on vertices are required by the hierarchical and adaptive visualization methods. Furthermore we need upper and lower bounds for data on elements, typically computable based on the error indicator values. Additional procedures are included to supply visualization methods with such information.

Particle tracing and related methods can be implemented similarly. For an efficient implementation we here have to process elements in the order in which they are traversed by an integral line of some vector field, e. g. a particle path. If we run the integration on level i, we therefore have to recover the "hierarchical history" of an adjacent i level element in each step, which can be done recursively. Here the hierarchical history is described as the corresponding list of *ELEMENT* structures and the recovering is done adjusting the list corresponding to the current element. This requires a procedure *neighbour()* on the coarse grid level and an evaluation of adjacency relations among the set of children of any parent element (cf. section 2). The latter can efficiently be done on general, hierachical grids taking into account local coordinate systems and the mapping from parent's local coordinates to local coordinates on a specific child and vice versa. These mappings only depend on the element type and the refinement rule and can therefore be precomputed. Details will be discussed in a forthcoming paper.

2 Economically stored nested grids

One of the main advantages of a procedural access to hierarchical data is that economically stored hierarchical grids can be addressed directly by the visualization. This allows us to handle even very large grid geometries interactively on a graphics workstation. Let us now sketch the minimal information, which is to be stored to address a hierarchical grid procedurally for numerical or graphical purposes.

For each element the geometry of its children is uniquely described by the refinement rule, if no additional grid alignment is taken into account. I. e. the coordinates of each $l + 1$ level vertex x^{l+1} in $\mathcal{N}(\mathcal{C}(E)) \setminus \mathcal{N}(E)$ can be evaluated as weighted sums over the coordinates of the l level parent vertices $x^l \in \mathcal{N}(E)$ with weights $\omega_{x^{l+1}}(x^l)$ depending solely on the refinement rule:

$$x^{l+1} = \sum_{x^l \in \mathcal{N}(E)} \omega_{x^{l+1}}(x^l) \, x^l$$

In general the number of refinement rules is small, such that element and vertex production rules, including the weights, can easily be stored in a lookup table. On curved boundary segments, where vertices generated by the refinement rules on the grid boundary are pushed afterwards onto the continuous smooth boundary, these new coordinates have to be stored additionally. We skip a detailed discussion here and only remark that the storage requirement is typically of lower order, because the boundary is a lower dimensional set.

MacroElement	
MacroNode	*macroNode
MacroElement	*neighbour
EcoElement	*self

EcoElement	
int	refrule
Node	**newNode
BoundPoint	*boundPoint
EcoElement	*child

Fig. 3. A sketch of possible user data structures which fully describe a hierarchical grid in a pseudo C notation (MacroNode and Node are the reference types for vertices on the coarse grid and new vertices generated during the refinement respectively. BoundPoint is the structure to store vertices on curved boundary segments)

A general hierarchical grid can be described by a list of macro elements, and the subdivision history for each of them, given by a tree of hierarchical elements. Therefore, in the user's application a macro element data structure *MacroElement* contains full information on the corresponding element: an identification of the nodes, including their coordinates or a reference to them, adjacency relations across element faces, and a reference to the tree of child elements. An economical hierarchical element data structure *EcoElement*, which corresponds to a node in any macro element's subdivision tree, consists of the index of the element's refinement rule, an array of identifiers for the new nodes generated by the refinement, to address data values on them, and finally a reference to an array of child elements (cf. Fig. 3). Let us mention that the highest level elements are not represented explicitly in the data base. Complete information on those elements is already present on their parent's level.

A typical visualization method now runs over the list of macro elements, and recursively processes higher level elements of the hierarchical tree. Thereby economically stored information for a child element is temporarily completed referring to the economical data structure for hierarchical elements, given as *EcoElement* data, and complete parent element information, which has already been stored in an *ELEMENT* structure (possibly slightly enlarged by some additional user data) in a preceding step of the recursion. Let us mention that besides the nodal coordinate vectors and the data references we can also generate information on element adjacency recursively, where we identify neighbouring cells of child elements as child elements of neighbouring cells.

Let us finally estimate storage requirements and thereby capabilities of a procedural approach, where we assume that only one refinement rule is used, which generates c children in each step. Now consider n levels of global refinement on a macro grid with m elements. Then the required storage for the hierarchy is

$$m \left(\frac{c^n - 1}{c - 1} \cdot EE + ME \right)$$

where EE, ME is the storage requirement for a single *EcoElement* or *MacroElement* structure respectively and furthermore $\frac{EE}{c-1}$ is an estimate for the storage needed per element on the finest grid level, if we neglect the small and constant memory block for the macro elements. A hierarchical, tetrahedral grid consisting of 10 million tetrahedrons on the finest level, where tetrahedrons are divided into eight children in each refinement step, with six new nodes per refinement (cf. Fig. 1), can be stored in $\sim$ 46 MB (here we suppose a need of 4 bytes in storage per integer, floating point number, and pointer). Compared to this, a non hierarchical storing of the finest grid level would require at least $\sim$ 340 MB (4 nodes and 4 adjacency references, plus $\sim \frac{1}{6}$ coordinate vector per element) without the chance to run hierarchical visualization methods.

3   A hierarchical and adaptive visualization strategy

Visualization methods, especially on 3D data sets such as isosurface extraction and color shading on slices, can benefit from the nested structure of the underlying grid. In the following we will focus on the isosurface case. Similar considerations hold for other visualization methods as well.

The cost to extract an isosurface from a given volume data set can be reduced enormously, taking into account hierarchical information. Instead of traversing all elements, like a standard marching cube strategy does, we can recursively test for intersections on coarser level elements E to decide whether the children $\mathcal{C}(E)$ have to be visited or not. If the considered function on the grid is smooth, this leads to a cost reduction of one order of magnitude up to a logarithmic factor. The intersection test on an element requires the calculation of robust data bounds. Simply taking into account the function values on the element vertices $x^l \in \mathcal{N}(E)$ will not be sufficient, because we might overlook information apparent on finer grid levels only, e. g. strongly curved segments of an isosurface (cf. Fig. 4). Let us suppose that we have at hand an estimate for the function's second derivatives — or something comparable — on each element of the hierarchy. Then a straightforward calculation of bounds is possible by means of Taylor expansion. In what follows we will see that such quantities are also essential to support an adaptive visualization strategy and can therefore with some care be reutilized here. For details we refer to [12].

Up to now, full information on the finest grid level is always extracted and visualized, disregarding that in areas where the isosurface is smooth, a considerably coarser resolution would visually be acceptable as well. But this would restrict the method's tree traversal to a coarser subtree and thereby considerably reduce the amount of graphic primitives which have to be rendered, and thereby substantially speed up image generation without missing fine details in other areas of the data sets. The locally coarser resolution can be obtained by

Fig. 4. On the left: isoline segments in 2D will be missed if just vertex information is taken into account to test for intersections (The same holds in 3D for isosurfaces), on the right: An adaptive traversal of a 2D grid leads to hanging nodes.

an adaptive stopping during the recursive traversal of the grid hierarchy on elements E, where some error indicator $\eta(E)$ is below a given threshold value ϵ. There is a variety of possible error indicators related to the local smoothness of the data (cf. [12] for a comparison). We found the jump of the normalized function gradient on the element faces to be an appropriate error criteria, because it measures angles between adjacent polygons, locally representing an isosurface. These jumps can be calculated and stored in a precomputing step on the vertices of the hierarchical grid, lying on the corresponding element faces. The error criteria $\eta(E)$ on an element E is then defined as the maximal indicator value on all its faces, i. e. on the vertices $x^{l+1} \in \mathcal{N}(\mathcal{C}(E)) \setminus \mathcal{N}(E)$. Following the adaptive strategy, hanging nodes (in 2D also called T–vertices) will in general be unavoidable. They occur on faces where we have a transition from coarser to finer elements, which are traversed by our adaptive algorithm (cf. Fig. 4). Hanging nodes lead to cracks in the isosurface. To rule out these artifacts we have to replace the true function value v on the finer elements at the considered face by interpolated values Iv coinciding with the function on the adjacent coarser element. The decision, whether to take the original function value, or the interpolation at a specific vertex, has to be based on the error indicators as well, i. e. we choose the interpolation at a vertex, if the stored indicator value is larger than ϵ. The interpolation is evaluated analogously to the calculation of coordinates for child vertices (see section 2). If we suppose the indicator values on coarser level vertices to be larger than indicator values on vertices appearing on the children of the corresponding elements, the resulting interpolation will be continuous and cracks will be ruled out (for bilinear faces some additional considerations have to be taken into account [12]). This is a natural assumption especially for fine grid levels and smooth data. In case of violations we adjust the indicator values, according to the condition, in an additional precomputing step. Let us emphazise that we operate only locally on single elements and thereby avoid an expensive non local construction of a conforming closure. The following algorithm sketches the hierarchical and adaptive isosurface extraction for an isosurface value α :

```
AdaptInspect(α,E) {
     Iv = Interpol(v, E);
     if IntersectionTest(α,Iv,E) {
          if (C(E) ≠ ∅) ∧ (η(E) > ε)
               for all Ẽ ∈ C(E)
                    AdaptInspect(Ẽ);
          else Extract(α,E);
     }
}
```

where *Interpol()* is the implementation of the above interpolation operator, *IntersectionTest()* checks whether α is contained in the image interval of the currently considered local function, and *Extract()* renders the local intersection of the isosurface with an element. Let us finally mention that our error indicator $\eta(E)$ can be utilized similar to a bound of the second derivatives to estimate higher order data contributions on a specific element. Thereby it allows a robust implementation of the intersection test without referring to additional information.

So far only the case of scalar data has been discussed. The presented concept can easily be generalized to vector valued data as well. Therefore we deal with the vector components seperately. Typically one might ask for isosurfaces of a scalar function of this vector data, i. e. the norm. Based on the component error indicator values an appropriate error indicator for this composite function can be computed. Furthermore the higher order offsets in the calculation of data bounds on elements depend on this composite error indicator.

4 Examples and Applications

To illustrate the performance of the hierarchical and adaptive strategy for different applications we now discuss some test cases. Fig. 5 depicts a significant example of an adaptive isosurface on a test data set which consists of precalculated values of an analytic function at the vertices of a tetrahedral grid. Fig. 6 deals with the density from a porous media calculation based on a hexahedral mesh (Numerical data provided by S. Oswald, Zürich). The hexahedrons are successively bisected, every time in one direction, cycling over x, y and z. The data is timedependent. Above, the isosurface for a fixed value is drawn at different times. They especially enlighten the adaptive strategy. Below a color shading of the density on an intersection plane is shown at the same times. Again black lines mark the intersected element faces. Function data at any time is interpolated based on a small set of time steps equipped with the corresponding data. Thereby the maximal error indicator of two time steps can be utilized as a possible error indicator for any interpolated data in between. This distinguishes our approach from the other efficient non hierarchical isosurface methods where a

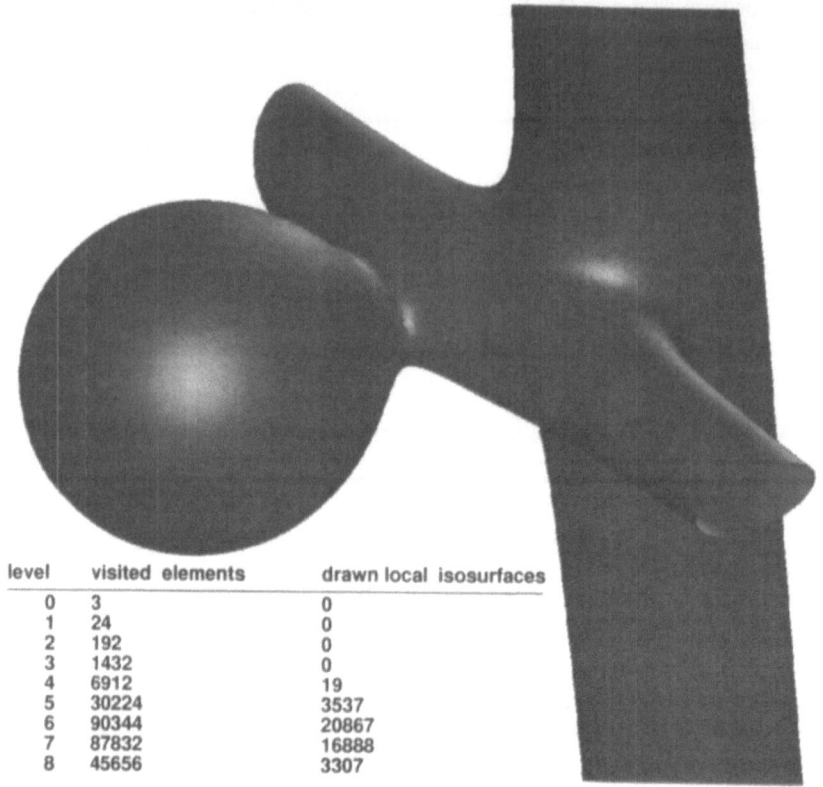

level	visited elements	drawn local isosurfaces
0	3	0
1	24	0
2	192	0
3	1432	0
4	6912	19
5	30224	3537
6	90344	20867
7	87832	16888
8	45656	3307

Fig. 5. An adaptive isosurface extracted from a 50 331 648 element data set. A table lists for each grid level the number of visited cells and the number of drawn local isosurfaces due to the adaptive stopping criteria.

preroll is unavoidable for each new time. In Fig. 7 we analyze the behaviour of the hierarchical and adaptive approach for an isosurface with a cusp type singularity on a sequence of successively refined tetrahedral grids. They are economically stored as described in section 2 and the final grid consists of ∼ 12 million elements. At each refinement step a tetrahedron is divided into eight child elements. A diagram shows in a logarithmic scale the total number of visited elements for different grid levels, including the coarser level elements passed by the hierarchical algorithm. We compare the results for a standard marching cube type algorithm, where we only count the elements on the finest level, with those for a solely hierarchical and for several hierarchical and adaptive runs of the algorithm, corresponding to different threshold values ϵ. This reflects the theoretically expected costs and underlines the considerable data reduction capability of the adaptive approach. Finally Fig. 8 shows a combination of several smoothly shaded, adaptive isosurfaces from the same data set and the result of an adaptive slicing algorithm on a non–conforming adaptive hexahedral grid on which a discrete solution of the Hamilton–Jacobi–Bellman equation is given

(Numerical data provided by L. Grüne [4], Augsburg). It demonstrates the applicability of our procedural approach to non standard grid geometries.

5 Conclusions

An efficient and general approach to visualize data from large scale scientific computations is presented. It especially takes into account the numerous types of data structures used in computation. These data types for hierarchical and adaptive meshes and function data fields are in general problem dependent. Furthermore they frequently allow the economical storing of minimal information to retrieve any local geometry and data needed in the computation. A mapping onto prescribed data formats is thereby often ruled out because of the resulting enormous storage requirements. The presented concept accesses mesh and function

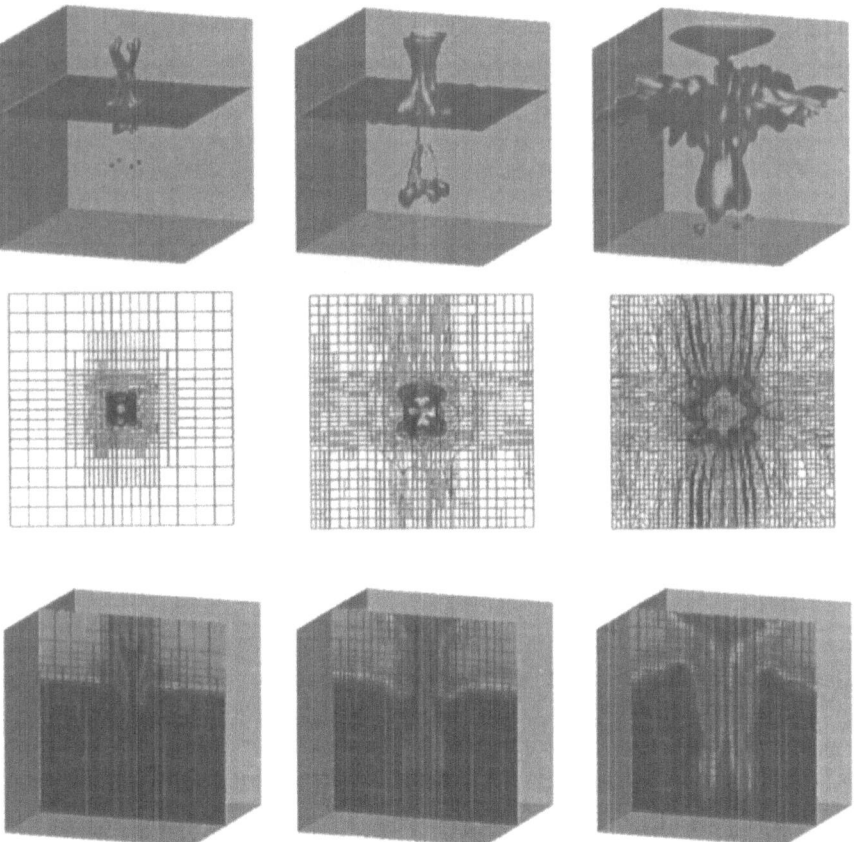

Fig. 6. At different times porous media data is visualized using isosurfaces (above) and color shading on slices (below). Additional black lines mark intersections with element faces in a projective view (in the middle) corresponding to the isosurfaces and on the intersection plane itself (below).

136

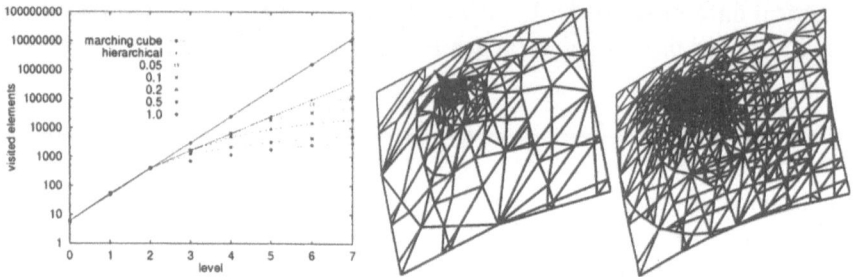

Fig. 7. In a diagram on the left we compare the proposed isosurface method for different threshold values ϵ with the standard marching cube type algorithm. On the right for two different ϵ grid models of the extracted isosurfaces are drawn.

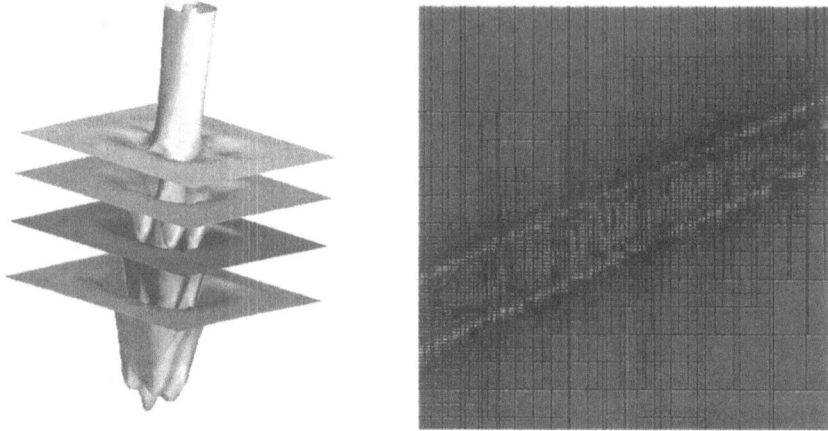

Fig. 8. Several adaptive isosurfaces from a porous media data set and and an adaptive slicing of a non–conforming hexahedral grid.

data only procedurally, temporarily completing information on single hierarchical elements. Thereby economically stored data on very large hierarchical grids can be handled. Hierarchical and adaptive visualization methods support an interactive data analysis. This generality of supported grid types and the required additional CPU load to supply local element information slightly increases rendering times compared to specialized methods on one specific grid type. Although the non economically stored, expanded data might fit into memory, the flexibility and the direct access to the user's data structures still compensate for this drawback. As interesting topics for future research we are currently considering the further optimization of particle tracing techniques, the implementation of volume rendering methods and the analysis of wavelet function spaces on arbitrary grids.

Acknowledgement

The authors thank W. Kinzelbach and S. Oswald from Zürich, who provided the porous media data set and L. Grüne from Augsburg for the non–conforming adaptive data set.

References

1. Cignoni, P., De Floriani, L., Montani, C., Puppo, E., Scopigno, R.: Multiresolution Modeling and Visualization of Volume Data based on Simplicial Complexes, Proceedings of the Visualization'95 , 19-26, 1995.
2. Dyer, D. S.: A dataflow toolkit for visualization, IEEE CG&A 10, No. 4, 60–69, 1990
3. Giles, M.; Haimes, R.: Advanced interactive visualization for CFD, Computing Systems in Engineering, 1(10): 51-62, 1990.
4. Grüne, L.: An adaptive grid scheme for the Hamilton-Jacobi-Bellman Equation, to appear in Numer. Math.
5. Haber, R. B.; Lucas, B.; Collins, N.: A data model for scientific visualization with provisions for regular and irregular grids, Proc. IEEE Visualization '91
6. Itoh, T.; Koyamada, K.: Isosurface Generation by Using Extrema Graphs, 77-83, 1994.
7. Lang, U.; Lang, R.; Rühle, R.: Integration of visualization and scientific calculation in a software system, Proc. IEEE Visualization '91
8. Laur, D.; Hanrahan, P.: Hierarchical Splatting: A Progressive Refinement Algorithm for Volume Rendering, ACM Computer Graphics 25 (4), 285-288, 1991.
9. Lucas, B.; et. al. : An architecture for a scientific visualization system, Proc. IEEE Visualization '92
10. Livnat, Y.; Shen, H. W.; Johnson, C. R.: A near optimal isosurface extraction algorithm using the span space, Transaction on Visualization and Computer Graphics, 2 (1), 73-83, 1996.
11. Lorensen, W.E.; Cline, H.E.: Marching Cubes: A High Resolution 3D Surface Construction Algorithm, ACM Computer Graphics 21 (4), 163-169, 1987.
12. Ohlberger, M.; Rumpf, M.: Hierarchical and adaptive visualization on nested grids, Mathematische Fakultät, Universität Freiburg, Preprint 22, 1996, to appear in Computing
13. Rumpf, M.; Schmidt, A.; Siebert, K. G.: Functions defining arbitrary meshes, a flexible interface between numerical data and visualization routines, Computer Graphics Forum 15 (2), 129-141, 1996.
14. Shen, H.-W.; Johnson, C.R.: Sweeping Simplices: A fast iso-surface extraction algorithm for unstructured grids, Proceedings of the Visualization'95 , 143-150, 1995.
15. Treinish, L. A.: Data structures and access software for scientific visualization, Computer Graphics 25, 104–118, 1991
16. Upson, C.; et. al.: The Application Visualization System: A computational environment for scientific visualization, IEEE CG&A 9, No. 4, 30–42, 1989
17. Wilhelms, J.;van Gelder, A.: Octrees for faster isosurface generation, ACM Trans. Graph. 11 (3), 201–227, 1992.

Editors' Note: see Appendix, p. 184 for colored figures of this paper

Simulation of Differential Interferometry and Comparison with Experimental Results

Rainer Wegenkittl and Eduard Gröller

Vienna University of Technology, Institute of Computer Graphics
Karlsplatz 13/186/2, A–1040 Vienna, Austria
e-mail: wegenkittl@cg.tuwien.ac.at and groeller@cg.tuwien.ac.at

Abstract This paper presents the computational simulation of a differential interferometer in a Mach-Zehnder arrangement with objects located outside the center of the interferometer. The computer simulation corresponds well to an experimental setup. This is illustrated for several basic phase objects. The theoretical models of these objects are discussed. The concept of an interactive visualization system for the analysis of phase objects is presented and finally some results allowing a comparison between experimental and simulated interferograms are shown.

1 Introduction

The visualization of computer simulated processes is an important field in computer graphics. Flight simulators, for example, are a well known example for such a task. Physical phenomena can also be computer simulated, allowing scientists to verify their simulation models with the results of an actual experiment. By doing this a mathematical model of a natural phenomenon can be adapted until it closely fits the result of the physical experiment. The use of computer graphics facilitates to decide how well the theoretical model corresponds to the physical phenomenon. Instead of comparing a huge amount of theoretically and experimentally acquired numerical data, the result of the simulation is visualized in a way similar to the result of the experiment. In this paper a system for the visualization of interferograms is presented. It shows, that the results of the simulated models agree nicely with the results of a real interferometer.

2 Interferometry

The non-intrusive density measurement within a fluid is a very important task of experimental fluid mechanics. One classical instrument for doing this job is the so-called *Mach-Zehnder Interferometer*. It utilizes the variation of the index of refraction of a homogeneous transparent medium according to its thermodynamic state [1]. Since the interferometer accumulates indices of refraction along a light beam passing through a medium, the resulting interferogram represents only the average variation of the index of refraction along the light beam. If an exact measurement of the index of refraction at each point in space is required, this method is only well-suited for planar flow fields. Such flows do not have a variation of the index of refraction along the light beam.

To determine the index of refraction at each point in space one can try to reconstruct the spatial field from the averaged field as it is provided by the Mach-Zehnder interferogram by means of costly inversion algorithms [7]. Another approach is to calculate theoretical Mach-Zehnder interferograms by performing the optical integration process numerically. Compared to the first method, the latter may sometimes be more appropriate, especially if the theoretical solution is given analytically.

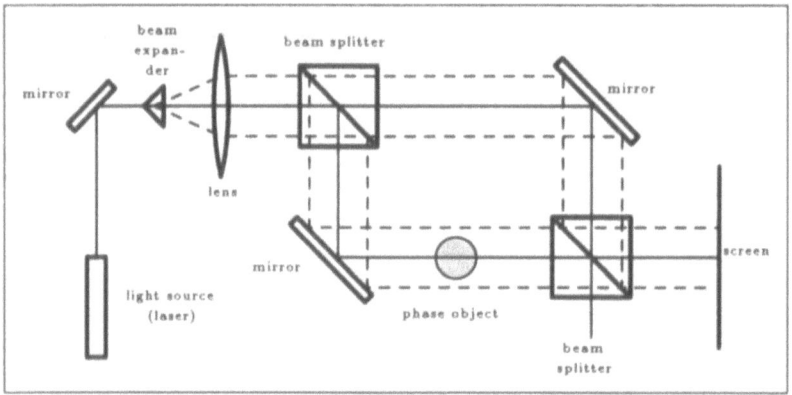

Figure 1: The concept of the Mach-Zehnder interferometer

Figure 1 shows the concept of the classical Mach-Zehnder interferometer. The light of a laser beam is expanded by a so-called beam expander. Then a lens produces a large parallel light beam, that runs into a beam splitter. The resulting two beams are mirrored into another beam splitter, where they are again combined to one beam. This arrangement has to be adjusted with high accuracy, so that the paths that both beams travel have exactly the same length. On the screen two identical and superimposed images appear . When a phase object is introduced in one of the two beams, one beam is shifted in phase due to the varying index of refraction. Thus phase extinction and amplification of the two superimposed images produce interference patterns.

There are two major drawbacks of the classical Mach-Zehnder interferometer. One is that the sensitivity of the system can not be adapted to the object under investigation. The other disadvantage is that the objects have to be positioned in the center of the interferometer represented by the beam splitters and two mirrors. The accurate adjustment of these parts is very difficult if they are far apart from each other, so only tiny objects can be investigated by the classical approach of interferometry.

3 Differential Interferometry

The above mentioned drawbacks can be avoided by using so-called differential interferometers. In Fig. 2 the experimental setup is shown. Again the laser beam is expanded to a large parallel beam which flows around the phase object under investigation. The object itself is now positioned outside of the interferometer. A lens focuses the beam onto two mirrors that are placed behind a beam splitter. Afterwards the beam is again combined by another beam splitter, it runs through another lens, and is finally projected onto a screen. A perfectly adjusted differential interferometer superimposes two identical images, so no interference effects can be seen in the resulting image on the screen.

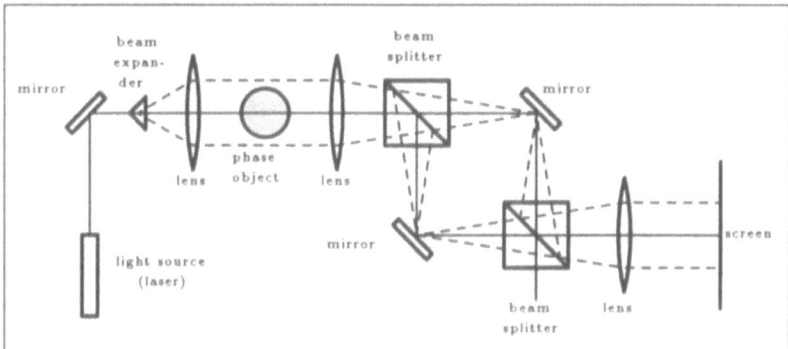

Figure 2: The concept of the differential interferometer

Now one of the two images is slightly shifted with respect to the other one. This is done by tilting one of the two mirrors of the interferometer. Thus an originally planar light wave is deformed by the object, divided into two identical parts and put together at the screen with a displacement δ. Now the two images will interfere and fringes will appear on the screen. If we assume, that the phase object causes a phase shift $h(y)$ in y-direction, the fringes are due to the differences $d(y) = h(y) - h(y - \delta)$. If δ is sufficiently small, $d(y)$ is proportional to the derivative of $h(y)$ with respect to y

$$\frac{d(y)}{\delta} = \frac{h(y) - h(y - \delta)}{y - (y - \delta)} \approx \frac{\mathrm{d}h(y)}{\mathrm{d}y} \ .$$

To superimpose a carrier fringe system, the second beam splitter is tilted slightly around an arbitrary axis as shown in Fig. 3. This causes a mutual inclination of the two corresponding light beams leaving the interferometer or, the projection F_1' of the focus F_1 onto the mirror is displaced from the focus F_2 by a distance ε. At each point on the screen, the two beams are inclined by an angle α. Thus a system of carrier fringes is produced, whose orientation is determined by the orientation of the tilt and the frequency is given by the size of ε. Since α will be different at different points on the screen, the fringes are not equidistant and parallel [6]. Depending on these tilt operations horizontal, vertical or diagonal fringes may be produced.

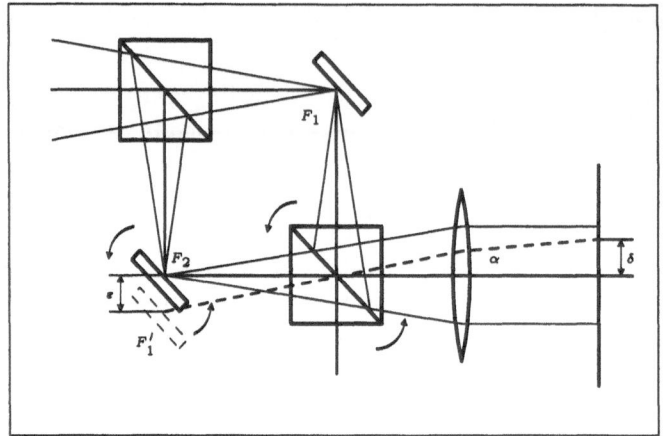

Figure 3: The principle of generating a carrier fringe system

4 Simulation and Experimental Verification

Two simple analytically defined models were integrated into a prototype software system. This prototype produces images of a simulated differential interferogram. The results have been compared to images that were produced by an experimental differential interferometer [2]. The images will be discussed in section 8. It turns out that the simulated models provide a good approximation for the given problems as the simulated interferograms closely resemble the experimental ones. In the following section the investigated models are described.

5 Model Data

Refraction data may be given analytically or numerically. Analytical data is given in one of three classes. The first class is made up of analytically defined objects whose index of refraction does not vary along the z-axis. An idealized flame whose width is assumed to not vary along the z-axis is an example for this class of systems.

The second class is made up of models, which analytically supply the indices of refraction at a plane behind the object. The plane is oriented to be perpendicular to the direction of the laser beam. In most cases these models represent cross sections of objects whose index of refraction does not change along the light beams. An example of this class is a cross section through a heated plate.

In the third class a formula for the index of refraction is given at each point within a three-dimensional space. These formulas often describe complex models whose calculation is a very time consuming task. To speed up the calculation of a differential interferogram, typically a specific program calculates the index of refraction for the specific model.

5.1 Flame Model

We will examine an idealized model of a flame [2], which means, that the radius of the flame is constant along its height. This assumption can be justified by the fact, that the width of a flame $a(z)$ (depending on the height z) can be modeled by $a(z) = a_0 + a_1 \cdot z$, where the condition $a_1/a_0 \ll 1$ has to hold. Thus the idealized flame is characterized by $a_1/a_0 = 0 \ll 1$. The next assumption is, that the index of refraction n is equal to a constant n_∞ (depending on the surrounding medium) everywhere except inside the test region, where it is assumed to be a function of the spatial coordinates.

So the formula for the temperature of the idealized flame can be derived from the general one

$$T(z,r) = \frac{T_0 \cdot a_0^2}{a(z)^2} \cdot e^{-\left(\frac{r}{a(z)}\right)^2} + T_u$$

which is simplified to

$$T(r) = T_0 \cdot e^{-\left(\frac{r}{a_0}\right)^2} + T_u \ .$$

The relation of the temperature to the index of refraction is given by

$$n(r) = G_w \cdot \frac{p_0}{R \cdot T(r)} + 1.0$$

where R denotes the gas constant (≈ 287.0), p_0 is the pressure of the atmosphere (≈ 100000.0), G_w stands for the Gladstone-Dale constant ($\approx 226 \cdot 10^{-6}$) and r denotes the distance from the center of the flame. Since the temperature depends on the distance from the middle of the flame, a coordinate transformation (by the distance to the middle of the flame) is done to integrate the indices of refraction along the laser light beam . The integral, that describes the phase shift for a laser light beam at a specific coordinate y is as follows

$$\frac{2 \cdot \Pi}{\lambda} \cdot \int_y^{a_v} (n(r) - n_0) \ dr \ ,$$

where λ denotes the wave length of the used laser ($\approx 660nm$) and n_0 shifts the refraction index outside the flame region to zero ($n_0 = 1.000288445585$).

5.2 Heated Plate Model

Another standard model describes a heated plate [2], which is vertical and also parallel to the laser light beam. The presented simulation model provides the index of refraction directly for every point on a plane perpendicular to the light beam. Thus no integration is needed, since the length of the heated plate produces only an additive constant to the indices of refraction.

Before calculating the temperature near the heated plate one has to evaluate the so-called similarity variable η by

$$\eta(y,z) = \sqrt[4]{\frac{g \cdot (T_w - T_\infty)}{4 \cdot v^2 \cdot T_\infty}} \cdot \frac{y}{\sqrt[4]{z}}$$

with g denoting the gravitational constant (≈ 9.81) and v the kinematic viscosity ($\approx 1.51 \cdot 10^{-5}$). The surrounding temperature is given by T_∞ and the maximum temperature by T_w. The relation of the real temperature to the similarity variable η has been investigated through several experiments and can be approximated by

$$T(y, z) = (e^{-0.8 \cdot \eta(y,z)}) \cdot (T_w - T_\infty) + T_\infty \ .$$

Again the index of refraction is derived from the temperature by

$$n(y, z) = G_w \cdot \frac{p_0}{R \cdot T(y, z)} + 1.0 \ .$$

Since the flame model does not vary along its z axis, shifting the simulated interferometer along that axis does not lead to any interfering effects. In comparison to that, the model of the heated plate allows much more testing of the simulation algorithm, as it produces interfering effects along both, the y and the z axis.

5.3   Numerical Simulations

Many simulation models can not be described by simple analytical formulas as mentioned above. Thus to make the system more versatile a two-dimensional or three-dimensional field of refraction indices can be imported to the simulation. This allows the use of different numerical packages and algorithms to numerically simulate the often complex and time consuming models. Interpolation between indices of refraction known at discrete spatial positions enables then the calculation of the differential interferogram.

In summary depending on the input data three different cases can be distinguished (Fig. 5):

- three-dimensional input data: The input array consists of a three-dimensional array of indices of refraction. An integral over the length of each light beam passing through the region under investigation has to be calculated.

- two-dimensional input data perpendicular to the light beam: This case resembles the case of the heated plate model. Each light beam penetrates the input data field at a single point, whose index of refraction is responsible for the phase shift of the light beam.

- two-dimensional input data parallel to the light beam: This case corresponds to the flame model, where in one direction (normal to the light beam) there is no variation of the index of refraction. Thus an integration along one axis of the input array has to be done for each light beam and the phase shift has to be calculated only for one scanline.

The data input due to numerical simulations is the most versatile one, whereas the other analytical models have been implemented primarily for testing.

6 Image Generation

Utilizing the above derived formulas, the phase shift $n_{y,z}$ of each laser light beam hitting the projection screen at (y, z) is calculated. Due to the tilting of the mirror in the interferometer, one of the two superimposed images is shifted along an arbitrary direction. This direction can be decomposed into a portion v_y in the y-direction and a portion v_z in z-direction. So the phase shift of the superimposed shifted image is $n_{(y-v_y),(z-v_z)}$.

To superimpose a carrier fringe system, the exit beam splitter of the interferometer has to be tilted around an arbitrary axis (see Fig. 3). This causes a mutual inclination of two corresponding laser light beams that exit the interferometer, thus the projection F_1' of focus F_1 onto the tilted mirror is displaced from the focus F_2 by a distance ε. On the screen, two corresponding laser light beams are inclined by an angle α, producing the desired effect of superimposed carrier fringes, whose orientation depends on the tilting axis of the exit beam splitter.

Unfortunately tilting the beam splitter causes a displacement δ of two corresponding laser light beams. To avoid this unwanted effect, the displacement can be extinguished by adding another displacement. This is done by tilting one of the mirrors of the interferometer. The tilting has no influence on the inclining angle α, so the carrier fringe system is not effected at all. The tilting of the exit beam splitter can be measured by two angles a_y and a_z, each one corresponding to either the y or the z axis. The formula for the phase difference of two superimposed laser light beams at a point (y, z) is given as

$$D_\varphi(y, z) = \frac{2 \cdot \pi}{\lambda} \cdot \left(n_{y,z} - n_{(y-v_y),(z-v_z)} + y \cdot a_y + z \cdot a_z \right)$$

If the two superimposing light beams are in phase at a specific point on the projection screen, they will appear as a bright point, whereas a phase difference of π produces extinguishing light beams, so the point on the projection screen remains dark. To calculate the brightness of the superimposed images at a specific point $P(y, z)$, one has to take the cosine (due to the assumed cosine shape of a light wave) of the phase difference. So the calculation of the differential interferogram is done by

$$P(y, z) = \cos \left(\frac{2 \cdot \pi}{\lambda} \cdot \left(n_{y,z} - n_{(y-v_y),(z-v_z)} + y \cdot a_y + z \cdot a_z \right) \right) \ .$$

6.1 Gray Scale Mapping

Since the evaluation of the above function provides the brightness for the superimposed images, this result can be directly used as an index into a color array [4]. In the experimental setup the projection screen is replaced by a black and white CCD-camera, whose output is captured and saved as a gray scale image. There is no use in trying to capture an image of a color camera, since the laser light is a monochromatic light source. Nevertheless some interferometers use a conventional white light source, producing colorful interferograms, where the color can be used to get more information on the object under investigation.

We use a simple linear gray scale with no gamma correction for displaying the interferograms [5]. Experiments showed that the output of the CCD is of low quality, so a gray scale array of 64 entries is sufficient to get a good approximation of the actual interferogram (see 8.1).

6.2 Aliasing

Due to the fact, that both, the theoretical and the experimental interferogram, have to be sampled at discrete points in space, severe aliasing artifacts [8][3] in regions of high phase-shift differences can occur. The pixels of a CCD camera have a certain extension, so spatial anti-aliasing is done automatically, because high and low light intensities cover the pixel and the CCD sensor supplies an averaged intensity. In the simulation process there are several possibilities to avoid aliasing effects.

One method is to use subsampling, where for each pixel on the screen several positions within the spatial extent of the screen pixel are calculated and the brightness of the screen pixel is the average brightness of the sampled locations. This resembles the process that happens within the CCD camera, but theoretically an infinity of light beams intersect each single pixel of the CCD sensor, so the actual subsampling is much better than the simulated one and subsampling is a time consuming task.

When simulating a differential interferometer, the actual phase differences at each pixel are known. With this knowledge one can determine specific regions where severe aliasing will occur by simply comparing the phase shift difference of one pixel with those of its neighboring pixels. If the difference between two neighboring pixels is greater than, for example, π, aliasing effects will occur for these pixels. With that knowledge, a mask can be calculated to depict those regions where severe aliasing effects occur (see Fig. 4). So a scientist can easily distinguish between fringes that are due to desired interfering effects and those effects, which are only due to aliasing.

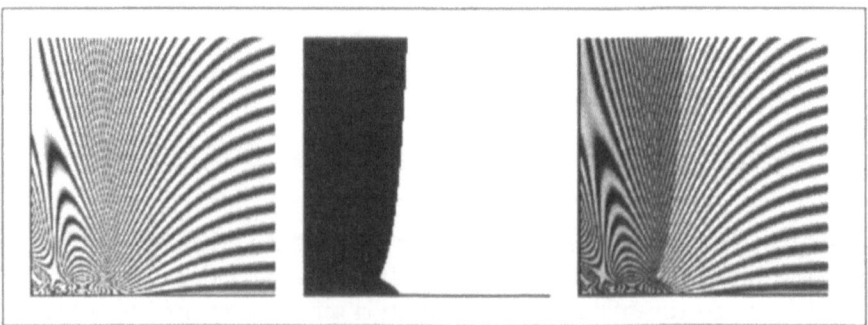

Figure 4: Aliasing effects, a mask for aliasing effects (threshold π), and the interferogram with superimposed mask

Aliasing artifacts can also be avoided by utilizing a property of the differential interferometer, namely the possibility to adjust the width of the carrier fringe system by changing the tilting of the exit beam splitter.

7 Interactive Software

The implementation of the test models showed good correspondence to the experimental results (see next section). To further increase usability we decided to implement a software system that allows a scientist to analyse input arrays of refraction indices interactively.

7.1 General Features

The software system is intended to allow a simple comparison of a mathematical model of temperature dissemination and the "real world" temperature dissemination. It allows the user to import a calculated density field, calculates a simulated interferogram out of this and gives the possibility to compare this interferogram with an image map taken from the CCD-camera of a real interferometer. Both images can be adjusted (scaled, moved and rotated) until they fit, and so differences between the simulation and the actual interferogram can be found easily. To detect what causes these differences, we implemented some additional visualization methods to illustrate the acquired density data.

7.2 Data Acquisition

To make the software system as versatile as possible, it has to handle various kinds of input data (scattered data, data on regular grids, data on irregular grids,...). To ease the data handling, we decided to resample the input data into an equidistant regular and rectangular grid. At the moment, this resampling is done by the nearest neighbor method, where the value for each point of the resampled grid is determined by the value of the nearest point of the input grid. This resampling can cause artifacts and so we plan to implement better interpolation methods for the resampling process. Nevertheless most input data is calculated on an equally spaced rectangular grid, so the resampling process leaves these grids unchanged if the extend of the resampled grid is chosen to be the same as the extend of the input grid.

The software system handles three different types of input data as shown in Fig. 5, the first case (top of Fig. 5) deals with a three dimensional field of input data. As described above this field has to be resampled into an equal spaced rectangular grid. Since the field is passed by the laser light beam in a fixed direction (which is parallel to one of the grid axis), the field can be reduced to a two dimensional field by accumulating (summing) all density values along the laser light beam. The resulting two dimensional field represents the phase shift (which is proportional to the accumulated density) of the laser light on every grid point on a plane perpendicular to the light beam. The interferometer shifts

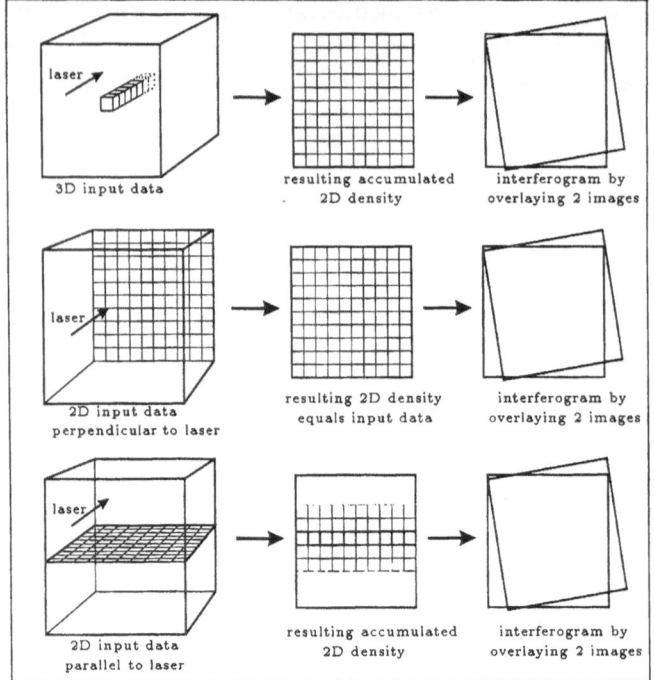

Figure 5: Three different types of possible input data

and rotates this plane and overlays the result with an unshifted version. Exactly the same is done by our simulation software. Each pixel of the density (phase shift) field is overlayed by a point of the shifted and rotated version of the field. Since the point that has to be overlayed does not lie exactly on a grid point, we have to interpolate linearly between neighboring grid points. The differences of the original phase shift and the shifted and rotated phase shift determine whether the resulting pixel is bright (both are in phase) or if the resulting pixel is dark (extinction by a phase shift of about π).

Some numerical simulations do not produce a three dimensional grid of densities, but a two dimensional grid which exactly represents the accumulated density values as described above (middle of Fig. 5). So the accumulation step can be skipped and the rest of the calculation process is exactly the same as before. The input data is overlayed by a shifted and rotated version of itself where again linear interpolation is used to determine the shifted and rotated density value. Differences in phase shift again determine the brightness of each pixel in the resulting interferogram.

The third method (bottom of Fig. 5) describes mathematical models that do not change along one axis (for example a model of an idealized flame, which has an infinite length). These models produce a two dimensional density field that is parallel to the laser light plane. The density values along the light beam have to be accumulated, and the result describes one single line of the resulting two dimensional density field. Since the field does not vary along one axis, this single line can be copied vertically to all other lines. So we have again a two dimensional field for the calculation of the interferogram.

7.3   Comparison of Simulated and Real Interferogram

After calculating the simulated interferogram, the simulation result has to be compared with a measured interferogram. During the simulation process some regions of the calculated interferogram may be distorted by interpolation arte-facts. Since these regions are of no interest for the comparision with a real inter-ferogram, they are masked. This mask can be calculated for different threshold levels to fit to user requirements.

To allow an easy comparision of simulated interferograms with a snapshot of a CCD camera in a real interferometer setup, a superimposing function is im-plemented in the software system. For a simulated interferogram a image file can be loaded and superimposed transparently. To match different settings of the CCD camera, both images, the simulated and the actual interferogram, can be scaled, shifted and rotated. Different colors for both images enable an easy distinction. By this superimposing function, differences in the interferograms, as for example differences in the width of the carrier fringes, can be detected and corrected by recalculating the interferogram. When size, position and spacing of the carrier fringes fit accurately, the superimposed bitmap allows a detection of regions, where the mathematical model is correct and regions, where the model differs from a real interferogram.

This comparison process could be automated by pattern recognition algorithms. As a future work we plan a system that automatically adapts this shift and rota-tion values for the simulated interferogram until it matches closely the measured interferogram. This reduces the sometimes tedious work of parameter tuning that has to be done by the user.

7.4   Data Exploration

To allow the scientist a verification of the density data that has been calculated, we implemented some data exploration tools. Depending on the dimension of the density array, different methods can be chosen. Multiple windows can be used, to compare different methods or one uses the same method with different settings. Each window has its own submenu to change specific parameters interactively or to save the contents of the window as a bitmap.

Data Exploration of 2D Density Fields. These methods do not distinguish, whether the data field is parallel or perpendicular to the laser light beam. The scientist has to take this into account when interpreting the data. A simple method is to color code the density and display it as a two dimensional bitmap. By clicking with the mouse into this bitmap, some parameters of the selected point, such as its position and refraction index (which is proportional to the den-sity), are shown beneath the cursor. Additional parameters like the maximum and minimum refraction index can also be superimposed in the window.

Another method is to view the two dimensional data set as a three dimensional height field. This height field can be visualized by either a point cloud, a grid, or a shaded surface. By clicking into the height field and dragging the mouse, the view can be rotated, allowing an inspection of the height field from different viewing angles. The shading of the surfaces is done with a single light source, which can be rotated around the height field to fit the lighting of the surface to user requirements.

Data Exploration of 3D Density Fields. To explore three dimensional data fields, we show a three dimensional point cloud representing the density field and a cutting plane, which can be rotated and translated interactively. In a second window the refraction indices within this cutting plane are shown as a color coded two dimensional bitmap.

The three dimensional visualization of the point cloud can be rotated around an arbitrary axis. To give a first impression of the density field, the transparency of each point is modulated by its corresponding refraction index. Higher density is mapped to higher transparency whereas lower density is mapped to a rather opaque point. When we look at the data set of a flame model, high temperature in the middle of the flame induce low density, thus points within the flame are visible, points that are farther away from the center of the flame are faded away by increasing their transparency.

By inserting a cut plane into the system, an exact measurement of density data can be done. Density values within the cut plane are calculated and visualized by a color coded bitmap which is displayed in a second window. By clicking with the mouse within the window, some information like the exact position of the selected point within three space and its corresponding refraction index are shown beneath the cursor.

Another visualization method for inspecting 3D density data is the use of isosurfaces. This methods builds up objects, whose faces are defined by all points of the same density. The marching cube algorithm is used to calculate isosurfaces. The density threshold can be adjusted interactively by the user.

7.5   Implementation

The described software system is intended to run on a low cost PC platform. We decided to implement it using OpenGL for doing the graphics and GLUT to do the window management. Since these tools are available for many platforms, the system is portable. On a 133 MHz Pentium PC and a grid resolution of 100x100x100 points the calculation of an interferogram takes about thirty seconds. Of course the calculation of interferograms of two dimensional input data is much faster.

8 Results and Conclusion

8.1 Flame Model

Figure 6 at the left shows a differential interferogram that was produced by an actual interferometer [2] whereas the right side shows the interferogram of a simulation. The object under investigation is a flame and the simulation has been done by using the equations of section 2.1. Fringes in horizontal direction appear due to a shifted exit beam splitter.

8.2 Heated Plate

Several possibilities of inducing a carrier fringe system are shown for the heated plate model (see Fig. 7–9). The chosen displacements are equal for all images and all directions and we have not suppressed any aliasing artifacts in these result images.

Figure 6: Actual and simulated differential interferogram of a flame

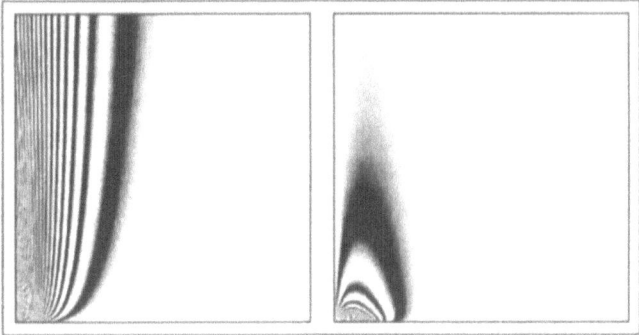

Figure 7: Interferogram with x (left) and y (right) displacement and no carrier fringes

The simulation of differential interferometry images allows the scientist to compare theoretical models with experimental results. A fast visual and interactive inspection of simulated interferometry images provides an effective tool to

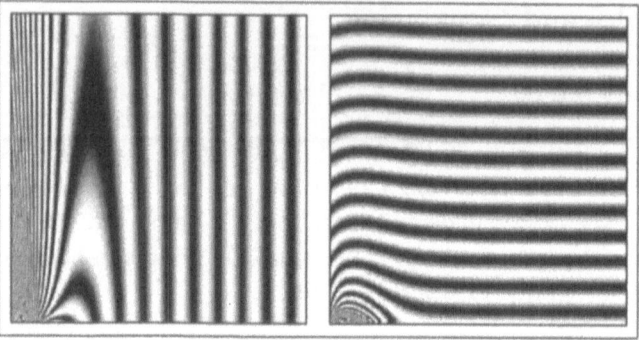

Figure 8: Interferogram with x (left) and y (right) displacement and vertical carrier fringes

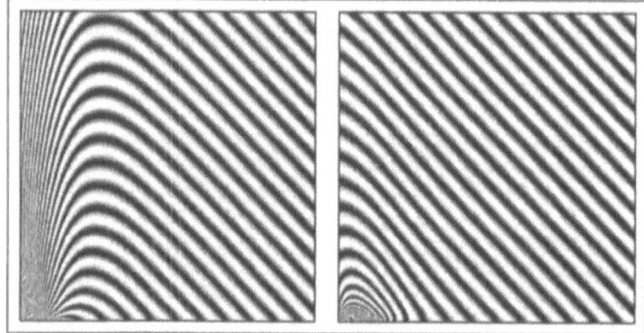

Figure 9: Interferogram with x (left) and y (right) displacement and diagonal carrier fringes

adjust theoretical models to natural phenomena. Once a theoretical model is established other visualization techniques might be used to get further insight into the theoretical model data.

9   Acknowledgement

We would like to thank Alfred Exner for supplying information about interferometers and for providing the results of experiments with a differential interferometer and Andreas Goldsteiner for helping in the software implementation.

References

[1] A. Dillmann. On the calculation of theoretical mach-zehnder interferograms from the given velocity potential of a cylindrical supersonic free jet. *Acta Mechanica*, 104:143–157, 1994.

[2] A. Exner. personal communication.

[3] J. D. Foley, A. v. Dam, S. K. Feiner, and J. F. Hughes. *Computer Graphics - Principles and Practice*. Addison-Wesley, 1990.

[4] S. Harrington. *Computergraphik - Einführung durch Programmierung*. McGraw-Hill Book Company, 1988.

[5] J. Pöpsel, U. Claussen, R.-D. Klein, and J. Plate. *Computergraphik - Algorithmen und Implementierung*. Springer Verlag, 1994.

[6] G. Pretzler, H. Jäger, and T. Neger. High-accuracy differential interferometry for the investigation of phase objects. *Engineering Optics, Meas. Sci. Technol*, 4:649–458, 1993.

[7] D. W. Sweeney and C. M. Vest. Reconstruction of three-dimensional refractive index fields from multidirectional data. *Appl. Optics*, 12:2649–2664, 1973.

[8] A. Watt. *3D Computer Graphics*. Addison-Wesley, 1993.

Hierarchical Streamarrows
for the Visualization of Dynamical Systems

Helwig Löffelmann, Lukas Mroz, and Eduard Gröller

{ `helwig` | `mroz` | `groeller` } `@cg.tuwien.ac.at`
Institute of Computer Graphics, Vienna University of Technology
Karlsplatz 13/186/2, A-1040 Wien, Austria
`http://www.cg.tuwien.ac.at/home/`

Abstract: Streamarrows are a technique to enhance the use of stream-surfaces by separating arrow-shaped portions from the remaining stream-surface. We present a hierarchical streamarrows algorithm as an extension to this technique: Streamarrows are locally chosen from a stack of scaled streamarrows textures to avoid too big or small streamarrows in the rendered image. We furthermore present techniques how stream-arrows can be extended into 3D, namely perpendicular to the stream-surface: streamarrows can be shifted slightly out of the streamsurface. Another extension in this category is to represent the outline of stream-arrows as 3D tubes. We show a set of images which have been rendered using this technique and report about ongoing research.

Keywords: Visualization, dynamical systems, streamsurfaces.

1 Introduction

One important area in the field of visualization [10] is the visualization of dynamical systems, e.g., flow fields. Various techniques and algorithms have been developed for such data [4].

A *dynamical system* represents the evolution of a set of variables over time [13]. Real world phenomena as, e.g., the stock market, ecological systems, or chemical reactions are often modeled or simulated as a dynamical system. Assuming a continuous evolution over time, a dynamical system is given as a set of ordinary differential equations (ODEs), which usually cannot be solved analytically. Visualization improves the exploration of such dynamical systems and promotes the understanding of the underlying dynamics.

Typically the temporal evolution of a dynamical system is depicted in *phase space*. Each coordinate of phase space is associated with a system variable. Thus each point in phase space represents a state of the system at a specific moment in time. The local evolution of a state can be expressed by a vector in phase space. This often results in a vector field representation of dynamical systems. The integration of such vector fields starting at a particular point in phase space produces the temporal evolution of the system. This temporal evolution is a curve which is called *trajectory*.

For quite a long time simple techniques like arrow plots or polyline representations of streamlines, streaklines, or pathlines have been used for visualizing dynamical systems [11]. Two-dimensional objects such as streamribbons [16] or streamsurfaces [5][15] are other important techniques for the visualization of dynamical systems. Furthermore extensions to these approaches such as stream-polygons and streamtubes [11] or the streamball technique [3] have been important contributions to the visualization of flow fields.

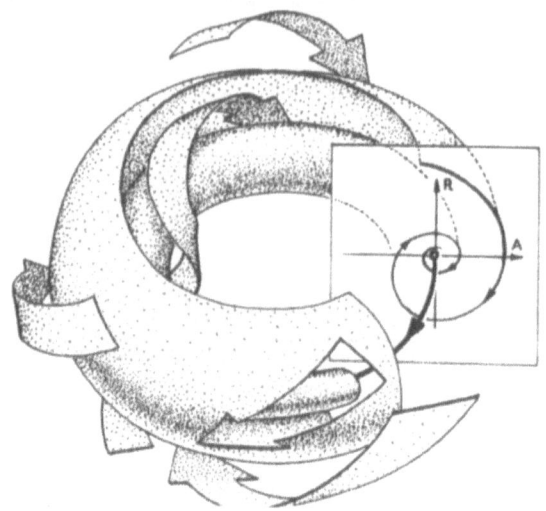

Fig. 1. Visualization of a dynamical system by using streamarrows [1]

Although the usage of streamsurfaces often improves the understanding of a system representation, occlusion becomes a problem whenever large opaque surfaces are used for visualization. An interesting technique to deal with this problem is utilized in the book by Abraham and Shaw [1]. They discuss various types of dynamical systems by using hand-drawn illustrations that represent the topological structure of these models. Streamsurfaces make up an important part of many of their images. To reduce the adverse effects caused by occlusion they use arrow-shaped parts of a streamsurface instead of the whole surface. In addition to that they use arrow-shaped holes within streamsurfaces to diminish occlusion. Refer to Figure 1 for a typical image of their book.

Other enhanced methods for the representation of large-scale surfaces have been discussed recently. Interrante et al., for example, derived a technique of representing a semi-transparent surface by using curvature-directed opaque strokes. Their approach is based on an analysis of artistic drawings [6]. In another paper Rheingans proposes an opacity-modulating texture for irregular surfaces as an attempt to reduce the problem of occlusion [12].

2 Streamarrows for mixed-mode oscillations

In cooperation with mathematicians, we investigated a class of dynamical systems that exhibit mixed-mode oscillations [8]. Such dynamical systems are characterized by a periodic pattern of alternating large and small oscillations.

We decided to use streamsurfaces to visualize the model. Due to the particular shape of streamsurfaces resulting from mixed-mode oscillations the resulting occlusion is a major problem. A typical streamsurface for the given dynamical system occludes important parts of itself by forming a shell-like roll with several turns. Self-occlusion is severe from almost any viewing position. Refer to Figure 6a to get an impression of a typical streamsurface within the investigated model. Throughout this project we investigated several techniques for highly occluding streamsurfaces [8]. Most of the methods have been inspired by the hand-drawn images of Shaw and Abraham [1].

The main extension to standard streamsurface techniques was the introduction of streamarrows. A streamsurface is thereby segmented into a set of arrow-shaped objects and the remaining surface portion. By assigning a certain level of semi-transparency to the streamarrows the viewer sees through the "holes" in the streamsurface and gets more information about the structure of the model. In addition to that, the use of streamarrows allows to visualize local information that is available on the streamsurface, e.g., direction of evolution, velocity, and local divergence or convergence (see Figure 6b).

The segmentation of the surface into streamarrows and the remaining surface portions was performed by mapping a regularly tiled texture of arrow-shaped patterns onto the streamsurface and cutting its patches along streamarrows borders. The texture was constructed by specifying a base tile, i.e. the shape of one streamarrow, and tesselating the texture using this base tile. Three sets of geometric objects, namely the inside of streamarrows, the outside, and the separating border, were extracted. After this segmentation either the stream-arrows or the remaining surface portions can be assigned a certain level of semi-transparency. See Figure 2 for an illustration of this method.

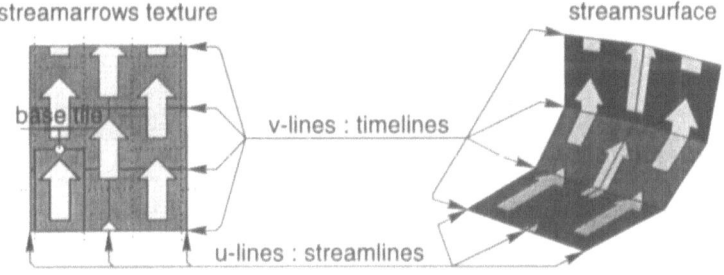

Fig. 2. Mapping the streamarrows texture to a streamsurface

A segmentation of a streamsurface into an entirely opaque portion and highly transparent holes (streamarrows) gives a good impression of the interior structure of a curved streamsurface while still retaining a good overview of the spatial arrangement of the streamsurface itself. Using homogeneously transparent surfaces would produce several layers of overlapping streamsurface segments which are quite difficult to interpret spatially.

3 Hierarchical streamarrows

The streamarrows approach [8] is based on a regular tiling of texture space. This is not well-suited for streamsurfaces that spread over regions of high divergence or convergence. The arrows become either too big or too small in certain areas. To eliminate this undesirable effect we present a hierarchical extension to the streamarrows technique.

3.1 A stack of streamarrows textures

The goal of our project was to develop an algorithm, which is capable of generating streamarrows that are almost equal-sized in the rendered image. Since we did not want to loose the ability to represent local divergence or convergence using the streamarrows technique, we decided to construct a hierarchical algorithm instead of a continuous solution. When flow, for example, diverges locally, our method switches discretely to the next detailed level of streamarrows. These streamarrows are smaller in texture space, but since they are used for divergent areas of the streamsurface, they are finally almost equal-sized to all other streamarrows in the rendered image. Refer to Figure 7 for an image which demonstrates this situation.

The *hierarchical streamarrows texture* is specified by the shape of one tile, i.e. the outline of an arrow, and two vectors dc and dr (see Figure 3) which define the offsets between adjacent columns and rows of arrows, respectively. Additionally there is a factor a which represents the scale relation between level

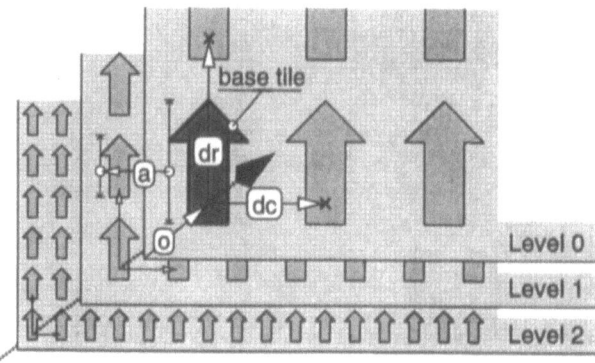

Fig. 3. Hierarchical streamarrows texture, specification parameters

i and $i+1$. If a=1/2, for example, the size of streamarrows is doubled, when the algorithm switches to the next coarser level. Finally there is a vector o, which is the offset of the entire texture with respect to the origin of texture space. Offset vector o becomes important, when animation is applied. Due to this specification each streamarrow can be addressed by exactly one ID given by three numbers. ID (*level*, *col*, *row*) identifies one streamarrow as a copy of the base tile, first translated by o+(dc · *col*, dr · *row*), and then scaled about the origin by a^{level}.

3.2 The separation algorithm

The new streamarrows technique itself is triangle oriented, because we produce our streamsurfaces as triangular meshes: the front of the streamsurface

($\approx$ timeline) is advanced through phase space while the streamsurface is integrated. Triangles are smaller in streamsurface areas, where curvature is high, flat areas of streamsurfaces are triangulated with relatively big triangles.

During the streamsurface algorithm vertices are assigned 2D texture coordinates. One coordinate of each vertex is set to the integration time of a streamline from the seed locus to the vertex. The other coordinate of each vertex is set in such a way that all vertices connected with one streamline get the same texture value, i.e. the 1D seed parameter of the start-point of the streamline.

To apply the hierarchical streamarrows texture to the streamsurface the hierarchical streamarrows algorithm processes the streamsurface triangle by triangle and performs the following separation algorithm:

```
activeTiles = {}              // . . . . IDs of active tiles
lockedTiles = {}              // . . . . IDs of locked tiles
FOR ALL Triangles tri DO:
| level:=findLevelOfTriangle(tri)  // . get most approprate level
| tiles:=getMaybeTiles(tri,level)  // . . . get overlapping tiles
| FOR ALL Tiles tile IN tiles DO:
| | IF NOT (tile.active OR tile.locked) THEN:
| | | IF overlap(tile,activeTiles) THEN: tile.lock
| | | ELSE:                              tile.activate
| intersect(tri,activeTiles)    // . . . . . do the separation
```

The algorithm needs two data structures in addition to the triangular mesh of the streamsurface: `activeTiles` stores the IDs of all tiles that actually are instanced within the streamsurface, whereas in `lockedTiles` all IDs of tiles are stored that overlap at least one active tile and thus should not be generated. Both data structures need to be be searched as fast as possible (in `overlap()` and `intersect()`) and easy to extend by a new tile. Since tiles are related via their 2D location in texture space and searching is also performed in a 'geographical' manner using texture coordinates (IDs), we use a linked data structure that closely represents the spatial relation between tiles.

The algorithm processes the streamsurface triangle by triangle. For each triangle `tri` the corresponding `level` in the hierarchical streamarrows texture is determined by comparing the size of triangle `tri` in texture space to its size in phase space coordinates. This ratio is used to find the most appropriate level in the stack of streamarrows textures. Then all tiles in that level, which might intersect triangle `tri` are determined (`getMaybeTiles()`). Tiles which are already activated or locked are omitted. All remaining tiles are checked, whether they overlap any active tile (`overlap()`). Tiles that overlap at least one active tile are locked (added to `lockedTiles`), and all the others are activated (added to `activeTiles`). See Figure 4 for an example with two triangles. After all tiles are checked triangle `tri` is intersected with all active tiles and separated into three sets: parts that belong to the arrows, parts that do not, and the separating outline. All three sets can be individually processed, e.g., assigned a certain level of semi-transparency, after the segmentation algorithm has finished.

The mechanism of `activeTiles` and `lockedTiles` ensures that neighboring triangles of a streamsurface are consistently covered by entire tiles of the hierarchical streamarrows texture. If the tile instanced for the currently processed triangle is overlapped by a tile which was already instanced for a previously processed and nearby triangle—this tile is active and of different level than the

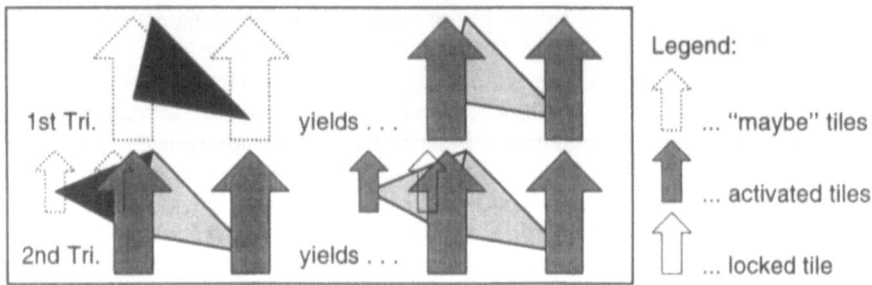

Fig. 4. Activating and locking tiles, example with two triangles

current tile—the acitve tile is used for the current triangle. This ensures consistency. The current tile which overlaps the active tile is furthermore locked.

4 Working with hierarchical streamarrows

After the separation of streamarrows has been performed (see section 3) various extensions can be applied. For example, two possibilities of extending streamarrows into 3D have been realized.

One possibility is to shift the separated streamarrows slightly in a direction perpendicular to the remaining streamsurface portions. To avoid any confusion— one could interpret the streamarrows as local solutions of the dynamical system, which is certainly not the case in most situations—the shifted streamarrows are connected with the remaining parts of the streamsurface by semi-transparent patches. 3D arrows are generated by this technique, which improves the perception of spatial location and orientation of the streamsurface (see Figure 8a). In this image streamarrows are furthermore enhanced by the use of anisotropic spot noise [14] showing stream- and timelines. Another 3D extension is the representation of the separating outline as 3D tube (see Figure 8b). Using this approach streamarrows can be realized without removing any part of the streamsurface. Nevertheless almost all of the advantages of the streamarrows technique are preserved as, e.g., indication of flow direction and local velocity.

Streamarrows are implemented in the scope of our DynSys3D system [7], which itself is developed on the basis of AVS [2]. Therefore animation is easily supported. AVS allows to export certain module parameters as input ports. Furthermore there are already some basic animation modules available as, e.g., *animated integer* or *animated float*. Connecting such a module to an exported module parameter is the simplest method of animating a visualization setup.

In our case several parameters can be animated to improve the visualization. Two examples are given: an animated offset vector o (see section 3.1 and Figure 3) can be used to move streamarrows along streamlines. Animating scaling factor a may be used to smoothly decrease the difference between arrows sizes throughout the visualization.

Our streamsurfaces technique has been enhanced by the capability of color coding local attributes as, e.g., velocity magnitude, integration time, divergence, or helicity [9]. This extension allows to focus the viewer's concentration on streamsurface regions that exhibit certain values of a local parameter.

5 Current developments

The project on streamarrows is still work in progress. Some of the ideas concerning this technique are currently under development. They are shortly discussed in this section.

3D Streamarrows can be seen as glyphs with a set of more or less free parameters. Scalar data such as velocity and helicity as well as vector data, for example, vorticity, may be mapped to a streamarrows parameter. We currently investigate the 3D shape attributes of streamarrows as glyph parameters, but also size, 2D shape, opacity, and color.

Up to now we can shift streamarrows slightly out of the streamsurface. This is quite an improvement in certain situations, but has its disadvantages in other cases. Especially if nearby streamsurfaces are not parallel at all—one streamsurface as a stable system solution can be imagined as such a case (see Figure 5)—this technique might yield wrong impressions. We therefore investigate

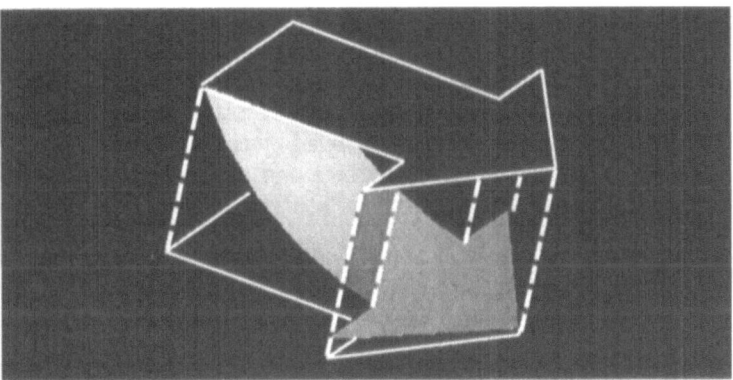

Fig. 5. Misleading shifted streamarrows (bright: integrated streamarrow)

another extension, which will allow to build up the shifted streamarrow as local solution by itself. Practically this is achieved by starting a new streamsurface at the shifted tail of our streamarrow and cutting the arrow out of this streamsurface. This technique will allow to represent the dynamical system not just at the streamsurface, but also in its vicinity. Other shapes of 3D streamarrows, e.g., a tent-like shape or variations of the 2D base shape, are also investigated.

Related to the hierarchical streamarrows texture we work on some other extensions concerning the placement of the streamarrows. One idea is to randomly position streamarrows on streamsurfaces. Another idea is to use local streamsurface attributes for the location of streamarrows. For example, surface curvature could be used for this purpose, since this has already been shown to be useful [6]. Finally we think of including viewing parameters to determine optimal placements of streamarrows. This idea was again inspired by the book of Abraham and Shaw [1], where these parameters are included in the hand-drawn illustrations as well.

6 Summary

Streamarrows are a useful technique to enhance the visualization of dynamical systems, i.e. flow fields. In this paper we discuss hierarchical streamarrows. A hierarchical texture is defined for streamsurfaces, which is given as a stack of different sized texture levels. Each level is specified as a set of arrow-shaped tiles that cover the texture of a given level in a regular manner. Controlling parameters of this streamarrows texture allows to improve the use of streamsurfaces by several means:

– Assigning a certain amount of semi-transparency to either the streamarrows or the remaining surface portions allows to reduce the problem of occlusion, which usually occurs when large scale surfaces are used for visualization.
– Local properties of the dynamical system are implicitly visualized by the use of streamarrows. Local convergence or divergence as well as flow velocity are directly represented by the shape of the arrows. Streamlines and timelines within the streamsurface are also visualized using this technique.
– The hierarchical approach allows to use the streamarrows technique also for ill-shaped streamsurfaces. Depending on the local area relation between texture space and phase space, the most appropriate level within the stack of streamarrows textures is chosen for streamarrows instantiation such that finally all streamarrows are more or less of equal size.
– Extensions into 3D furthermore enhance the streamarrows technique. Three-dimensional streamarrows can be modeled by shifting the separated streamarrows slightly above the streamsurface and connecting them with the remaining surface portions by the use of semi-transparent patches. If no parts of the streamsurface should be removed, the arrow outlines can be modeled as 3D tubes.

Currently the following concepts are integrated into our system[1]:

– Interpreting a streamarrow as a glyph allows to further extend this technique. Local attributes of the dynamical system as, e.g., helicity, can be mapped to glyph parameters and thus visualized.
– Using 3D streamarrows it is possible to visualize the vicinity of a streamsurface. Convergence or divergence perpendicular to the streamsurface will thus be visualized.
– Arrow placement is another topic of current research. Breaking the regular distribution of streamarrows within a level allows to selectively place arrows at places of special interest.

Streamarrows are a powerful technique for the visualization of dynamical systems. Streamarrows and hierarchical streamarrows enhance the use of streamsurfaces. These approaches allow an easier interpretation of complex, e.g., curly streamsurfaces. Streamarrows are of limited use for highly convergent or divergent streamsurfaces. For such cases hierarchical streamarrows as proposed in this paper are a better choice. Comparing images produced with the streamarrows technique, e.g., Figure 9, to those in the book of Abraham and Shaw [1], for example, Figure 1, shows that this approach is capable of generating results that are similar to impressive hand-drawings of an artist.

---

[1] See also a web-page compiling results via the following URL:
http://www.cg.tuwien.ac.at/research/vis/dynsys/HierStreamarrows97/

References

1. R. H. Abraham and C. D. Shaw. *Dynamics - The Geometry of Behavior.* Addison-Wesley, 1992.
2. Advanced Visualization System Inc. *AVS Developers Guide, Release 4,* May 1992.
3. M. Brill, W. Djatschin, M. Hagen, S. V. Klimenko, and H.-C. Rodrian. Streamball techniques for flow visualization. In *Proceedings Visualization '94,* pages 225–231, October 1994.
4. E. Gröller. Application of visualization techniques to complex and chaotic dynamical systems. In M. Göbel and H. Müller, editors, *Visualization in Scientific Computing,* pages 63–71. Springer, 1995.
5. J. P. M. Hultquist. Constructing stream surfaces in steady 3d vector fields. In *Proceedings Visualization '92,* pages 171–177, October 1992.
6. V. Interrante, H. Fuchs, and S. Pizer. Illustrating transparent surfaces with curvature-directed strokes. In *Proceedings Visualisation '96,* pages 211–218, 1996.
7. H. Löffelmann and E. Gröller. DynSys3D: A workbench for developing advanced visualization techniques in the field of three-dimensional dynamical systems. In N. Thalmann and V. Skala, editors, *Proceedings of The Fifth International Conference in Central Europe on Computer Graphics and Visualization '97 (WSCG '97),* pages 301–310, Plzen, Czech Republic, February 1997.
8. H. Löffelmann, L. Mroz, E. Gröller, and W. Purgathofer. Streamarrows: Enhancing the use of streamsurfaces for the visualization of dynamical systems. To be published in the Journal 'The Visual Computer', 1997.
9. H. Löffelmann, Z. Szalavári, and E. Gröller. Local analysis of dynamical systems – concepts and interpretation. In N. Thalmann and V. Skala, editors, *Proceedings of The Fourth International Conference in Central Europe on Computer Graphics and Visualization '96 (WSCG '96),* pages 170–180, Plzen, Czech Republic, February 1996.
10. G. M. Nielson and B. Shriver. *Visualization in Scientific Computing.* IEEE Computer Society Press, 1990.
11. F. H. Post and J. J. van Wijk. Visual representation of vector fields: Recent developments and research directions. In Rosenblum et al., editor, *Scientific Visualization - Advances and Challenges,* pages 367–390. Academic Press, 1994.
12. P. Rheingans. Opacity-modulating triangular textures for irregular surfaces. In *Proceedings Visualisation '96,* pages 219–225, 1996.
13. A. A. Tsonis. *Chaos - From Theory to Applications.* Plenum Press, 1992.
14. J. J. van Wijk. Spot noise. In *Proceedings SIGGRAPH '91,* pages 309–318, July 1991.
15. J. J. van Wijk. Implicit stream surfaces. In *Proceedings Visualization '93,* pages 245–252, 1993.
16. G. Volpe. Streamlines and streamribbons in aerodynamics. In *27th Aerospace Sciences Meeting, Reno, NV,* January 1989. AIAA Paper 89–0140.

Editors' Note: see Appendix, p. 185 f. for colored figures of this paper

Parametrizable Cameras for 3D Computational Steering

Jurriaan D. Mulder[1] and Jarke J. van Wijk[2]

[1] Centre for Mathematics and Computer Science CWI
P.O. Box 94079, 1090 GB Amsterdam, the Netherlands
mullie@cwi.nl
[2] Netherlands Energy Research Foundation ECN
P.O. Box 1, 1755 ZG Petten, the Netherlands
vanwijk@ecn.nl

Abstract. We present a method for the definition of multiple views in 3D interfaces for computational steering. The method uses the concept of a point-based parametrizable camera object. This concept enables a user to create and configure multiple views on his custom 3D interface in an intuitive graphical manner. Each view can be coupled to objects present in the interface, parametrized to (simulation) data, or adjusted through direct manipulation or user defined camera controls. Although our focus is on 3D interfaces for computational steering we think that the concept is valuable for many other 3D graphics applications as well.

Introduction

Computational steering allows a researcher to change parameters of a running simulation and immediately receive feedback on the effect of these changes. As a result, the researcher can gain a much better insight in the behavior of the simulation, he can correct erroneous values for input, and he can steer the simulation into a desired direction. At CWI an environment for computational steering has been developed that allows a researcher to construct and use customized 2D and 3D user interfaces for his simulations. These interfaces consist of both the visualization of the simulation output, and the input widgets that allow the researcher to manipulate the input parameters of the simulation.

The definition of the view, i.e. the transformation of 3D scenes to a 2D screen, is a crucial aspect in 3D graphics applications. This is particularly the case in 3D user interfaces for computational steering. First of all, improper viewing of a 3D scene can prevent a good understanding of the 3D scene; important information can be withheld from the user which leads to an incomplete or, even worse, an incorrect interpretation of the scene. Secondly, in computational steering the 3D scene is dynamic. It is not known in advance how the simulation will evolve and therefore it is not known in advance what the 3D scene will look like. This implies that the viewing also must be dynamic and easily reconfigurable Thirdly, in computational steering applications the user must be able to interact with the scene. Therefore, the user must have an adequate view of the objects to manipulate and be enabled to get good visual feedback of the changes in the scene which result from his actions. And finally, because he creates his own custom 3D user interface to a simulation, the user has control over the layout of the interface and therefore he must be enabled to specify the view on this interface tailored to his needs.

In this paper, we present a method for the definition of views in custom 3D user interfaces for 3D computational steering applications. The method is based on point-based parametrizable cameras. A view is created by placing a camera in the 3D scene. The characteristics of the camera such as its position and orientation define the view, which is presented to the user in a separate window. These characteristics are defined by the position of the camera's control points. The control points can be moved by the user, parametrized to simulation data, or coupled to objects present in the scene. Although we have developed this method to be used within our computational steering environment we believe that it can easily be adapted for usage in many other 3D graphics applications and software packages.

In the next section we describe related work on viewing methods for 3D scenes (Sect. 1.1) along with a brief description of the computational steering environment developed at CWI, including the tool developed to create custom 3D interfaces (Sect. 1.2). Section 2 describes the requirements of view definitions in 3D interfaces for computational steering applications followed by a description of the point-based parametrizable camera object (Sect. 3). In Sect. 4 some example applications of the camera object are shown. Finally, Sect. 5 gives some concluding remarks and indicates areas of future research.

1 Related Work

1.1 Viewing Control

Several projection methods exist for displaying a 3D scene on a 2D display device [4]. Most applications however, use the perspective projection model where the view is defined by specifying the center of projection, the view plane, and the clipping volume. On top of this model, different view manipulation metaphors have been developed for both user controlled and automated view manipulations.

Ware et al. defined the *eyeball in hand*, *scene in hand*, and *flying vehicle control* metaphors for exploration and virtual camera control in virtual environments using a six degree of freedom input device [16]. They found that the different metaphors each have their advantages and disadvantages depending on the particular task that is to be performed. The flying vehicle metaphor for instance was more useful in navigating through an interior while the scene in hand metaphor was useful for manipulating closed objects.

Phillips et al. presented a method for automatic viewing control to support 3D direct manipulation techniques of objects in a scene [12]. Through automatic viewing adjustments their system tries to avoid viewing obstructions and reduce the problems with degenerate axes common to most direct manipulation techniques.

Gleicher et al. presented the concept of *through-the-lens camera control* [5] where the user can manipulate a virtual camera by controlling and constraining features in the image seen through its lens. Constrained optimization is used to compute the time derivates of the actual viewing parameters which allows for controls to be defined independently of the underlying view parametrization.

Many other researchers have addressed research issues associated with view definitions and in particular viewpoint movements [7,14,8,2,3,6,11]. However, most of the developed techniques are heavily application dependent. They do not allow a user to interactively define multiple views on a 3D scene in an intuitive, graphical manner such that each view can be configured and parametrized according to the user's needs.

1.2   The CSE and PGO editor

The Computational Steering Environment (CSE) which is being developed at CWI [15], is based on two major concepts: A central *Data Manager* surrounded by processes called *satellites*. The Data Manager takes care of centralized data storage and event notification, and the satellites can connect to and communicate with the Data Manager by the use of a 'publish and subscribe' paradigm, see Fig. 1. Typically, one of these satellite processes is a simulation. Via a few simple function calls the simulation can open a connection to the Data Manager, connect the appropriate input and output variables, and update and retrieve their values when needed. Once the input and output variables of the simulation have been connected to the Data Manager, additional visualization and data manipulation satellites can be used to visualize the simulation results and steer the simulation by altering the input parameters present in the Data Manager. Several general satellites have been developed for this purpose, such as a logging satellite, a calculator satellite, and a slicing satellite. The most important satellite however is the *PGO editor* [9,10].

The PGO editor is a tool that enables the user to create 2D and 3D user interfaces for the visualization and manipulation of the (simulation) data present in the Data Manager. These interfaces are constructed out of *point-based Parametrized Geometric Objects*. A set of simple basic objects (such as a sphere, a line, and a box) is provided to the user out of which he can construct complex, composite input/output widgets. The geometry of these basic objects is defined with control points. The control points indicate important perceptual characteristics of the objects such as 'tip', 'bottom', 'corner', etc.. Changes of the positions of these control points change the geometry of the objects. Therefore, the geometry of the objects can be parametrized to data in the Data Manager by parametrizing the point positions to the data. This is accomplished by assigning a domain of allowed positions to a point and using Cartesian or spherical coordinates to describe the point's position inside this domain. These x, y, and z, or *azimuth, elevation,* and *radius* parameters are then linked to variables in the Data Manager. As a result, the user can manipulate the simulation input parameters present in the Data Manager by dragging the control points, i.e. the objects can be used for user input through direct manipulation, while changes

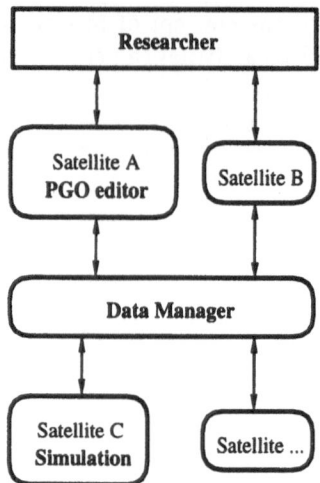

Fig. 1. Architecture of the Computational Steering Environment.

in the simulations output data cause the objects in the interface to transform, i.e. the objects can be used to visualize the output data. Further, to construct complex, composite input/output widgets hierarchical inter-point connections can be established to propagate control point transformations from one point to another and create inter-object relations.

Figure 2 shows a simple example of an input/output widget that is constructed in the PGO editor. The arrow consists of two basic objects (a cylinder and a cone) that are each defined with three control points. One control point is parametrized to variables in the Data Manager such that the length and orientation of the arrow can be altered.

2 Viewing in Computational Steering Applications

The viewing operations to be provided in 3D computational steering interfaces must enable the user to

- Define multiple views on the 3D scene.
  The benefits of multiple views are obvious: different viewing configurations that can be displayed simultaneously can significantly improve the insight in a 3D scene and relieves the user from having to redefine or switch between different views.
- Parametrize a view to (simulation) data.
  For automated view control the parametrization of the view to data is required. This allows for automated walk-throughs in complex visualizations calculated by an external process (a satellite in the CSE), or coupling of a view configuration to simulation output data.

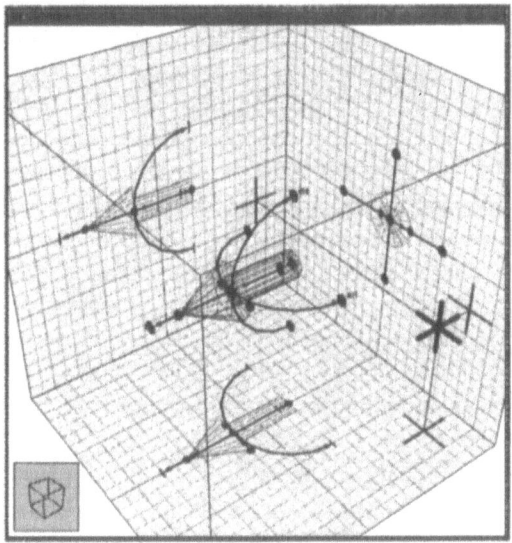

Fig. 2. An arrow defined in the PGO editor.

- Couple a view to entities in the interface.
  This allows the user to automatically trace objects as they move through the 3D scene. This is important if the user wants to focus on one particular aspect of the object or manipulate (part of) the object while it is moving, which is a task that can be quite difficult to perform if the object is moving fast or is located in complex, crowded scenes.
- Construct user defined controls over the view and change a view by direct manipulation.
  The user must also be able to configure a view that he can control and manipulate himself while using the interface to steer a simulation. The manipulation mode however, can differ from application to application and from view to view. Sometimes the user may want to only translate a view along a particular path while on other occasions he might want to be able to rotate the view about a particular axis. Therefore, the user must be enabled to define different view manipulation modes that control the view according to his needs.

To meet these requirements we have developed the point-based parametrizable camera object. This object is provided in the PGO editor and allows for easy definition of different view configurations in the user defined 3D interface.

3 Point-Based Camera Object

The PGO editor provides one main view on the scene. This view is used as an overview of the entire scene. The user can change this overview by direct manipulation of a small box icon present in the main view. To define additional views in the PGO editor a camera metaphor is used. This metaphor allows for easy and intuitive definitions of different view configurations. To create a view a camera object is positioned in the 3D scene. The geometric properties of the camera object are defined by three control points. The camera's position and direction of view can be defined in a very natural way by the use of these control points. One control point is used to define the camera's position and the second control point defines the direction of view; the camera is pointed towards this control point. The third control point is used to define the camera's roll, i.e. the rotation about the line of sight. The plane that is defined by the three camera control points is the median plane of the camera. An important attribute of the camera is the diameter of the lens opening: this defines the zoom factor. The diameter of the lens is derived from the distance between the third control point and the line of sight.

Because the camera object is point-based, it is fully integrated in the PGO editor. The same flexibility that is provided for the geometric objects applies to the camera, and the same interface and manipulation techniques are used for both the construction of the 3D scene and the configuration of the cameras. Each instance of the camera object displays the scene from its point of view in a separate window. The user can edit or manipulate the 3D scene in the main view or in one of the camera windows.

Different viewing configurations can be established by connecting the camera's control points to the control points of other objects, sharing the control points with other objects, or parametrize the control points to variables in the Data Manager. The user can also create custom camera control widgets out of the geometric objects or change the camera configuration by direct manipulation of the camera's control points.

Figure 3 shows the point-based camera object in one of many possible configurations. The camera is displayed in wire-frame mode to reveal the center control point. The camera's position is parametrized to the variables c_x, c_y, and c_z, its direction of view to c_ele and c_azi, and the roll and zoom to c_roll and c_zoom.

4 Examples

Figure 4 depicts a robot arm that was constructed with the PGO editor. The robot arm comprises several different rotational and translational joints. The user can control the robot arm with the sliders, where each slider manipulates one joint, or by manipulation of the robot arm itself.

Two camera objects are used in this scene. One camera is rigidly mounted on the end effector of the arm. Therefore, this camera moves with the end effector and always displays what the end effector 'sees'. In front of this camera a slider has been constructed to control the zoom of the camera. This slider always remains in front of the camera such that the user can adjust the zoom factor by manipulating the slider in the camera window. The other camera has a static

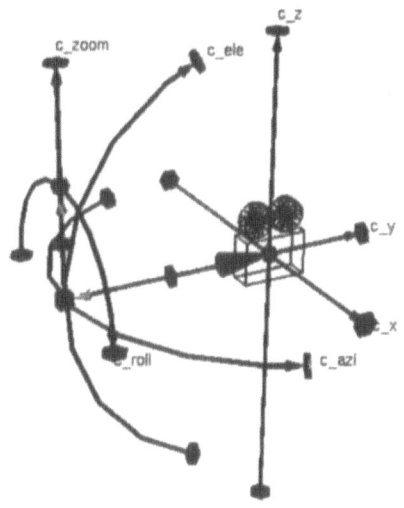

Fig. 3. Point-based camera.

Fig. 4. Robot arm with camera objects.

position. Its direction of view however, is coupled to the robot arm's end effector. So, this camera traces the end effector from a fixed position. The views of both cameras are depicted in Fig. 5.

Fig. 5. View from the cameras present in the robot arm interface. Left the view from the camera mounted on the end effector and right the view from the camera that traces the end effector.

The picture on the left in Fig. 6 shows an interface to a path planning application developed by K. Trovato [13] and L. Dorst et al. [1][1]. The interface shows two representations of a car parking problem. One representation visualizes the task space: the street, the car, and the two obstacles in between which the car has to be parked by the path planning program. The other representation shows the configuration space, called c-space. Here the three parameters that describe the configuration of the car are visualized: two position parameters x and y, and the car's orientation ϕ. Each plane represents a particular angle of the car. Because ϕ is cyclic the planes are ordered in a cyclic order. Each plane represents the x and y parameters of the car. The areas in the planes that are not filled represent a parameter configuration for the car such that it does not interfere with the obstacles. The user can examine the c-space by manipulating two boundary planes in the c-space visualization to select a region to be visualized.

The car can be dragged to a new initial or goal position by manipulating the car itself or its representation in the configuration space: a small sphere. Also, the two obstacles can be resized by direct manipulation. The traveled path of the car is visualized with wire frame projections of the car in the task space, and with a polyline in the configuration space. A camera has been mounted on top of the car to get an impression of the scene from the driver's perspective as shown in the picture on the right in Fig. 6.

---

[1] The path planning software is ©by Philips Laboratories, 1988. Philips has four patents pending related to the vehicle planning methods and control.

Fig. 6. Interface to a path planning application. On the left an overview of the interface, on the right the scene from the driver's perspective.

Figure 7 illustrates a different use of the camera object. Here, the control points of the camera have been parametrized to variables in the Data Manager. A separate satellite process is used to control these parameters and thereby to steer the camera. The satellite process has a user interface with several buttons that allow the user to move or rotate the camera and zoom in or out. The satellite process interprets the user's actions on the buttons, computes new values for the variables to which the camera's control points are parametrized and stores these in the Data Manager. The control points of the camera then adapt their position according to the new values and thereby alter the view.

An extensive library of such satellite processes could be developed where each satellite controls the camera in a different fashion. The user can then select one of these satellites according to which is needed for the particular application. The use of a satellite process to control the camera motions via variables in the Data Manager also enables the usage of special input devices such as a head tracking device or a Spaceball. The satellite process is then used as a device driver that reads out the device's parameters and stores them in the Data Manager.

These examples illustrate some possible applications of point-based parametrizable cameras in the PGO editor. The camera object allows the user to fully customize the viewing of a 3D user interface in the same manner as the user has constructed that interface itself. Many different viewing configurations can be realized yet the camera model remains easy and intuitive to use.

5 Conclusion

3D Graphics applications can greatly benefit from multiple user defined or automated views. This holds particularly for customized 3D interfaces for computational steering where the user is to interact with a dynamic 3D scene. Different

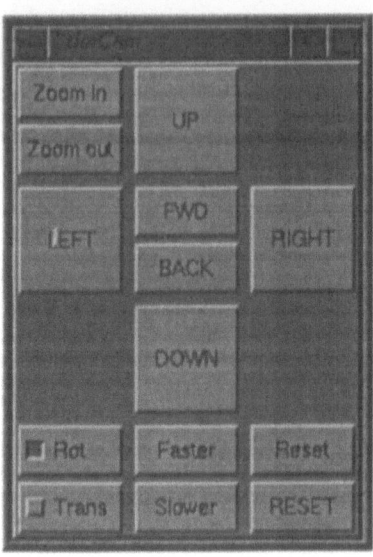

Fig. 7. A separate satellite to control camera movements.

viewing configurations will aid the user in understanding the 3D scene and provide support for manipulations on the objects in the scene. The user should be enabled to create and define the views as he also creates the interface itself and therefore will have specific ideas for the views on the interface.

The point-based parametrizable camera object presented in this paper provides an easy and intuitive method to define multiple views with different configurations. The user can easily manipulate the views themselves, link the views to objects in the scene or parametrize the views to (simulation) data for automated camera control. Even though we have developed the camera object for usage in the PGO editor we think that the principle can easily be adapted for usage in other applications.

One of the research items we want to explore in the near future is the aspect of view selection. If multiple views are defined on a 3D scene, can we develop techniques for automated selection of the best view for a particular task? For instance, if multiple cameras are used to trace different objects through the 3D scene it might be useful to automatically pop up the appropriate camera view upon selection of the object the camera traces. Other future research areas include the use of multiple views in virtual reality applications and the development of techniques for automated camera generation.

References

1. L. Dorst, I. Mandhyan, and K. Trovato. The geometrical representation of path planning problems. *Robotics and Autonomous Systems*, 7:181–195, 1991.

2. S.M. Drucker, T.A. Galyean, and D. Zeltzer. CINEMA: A system for procedural camera movements. In D. Zeltzer, editor, *Computer Graphics (1992 Symposium on Interactive 3D Graphics)*, pages 67–70, 1992.

3. S.M. Drucker and D. Zeltzer. CamDroid: A system for implementing intelligent camera control. In P. Hanrahan and J. Winget, editors, *1995 Symposium on Interactive 3D Graphics*, pages 139–144, 1995.

4. J. Foley, A. van Dam, S. Feiner, and J. Hughes. *Computer Graphics: Principles and Practice*. Addison-Wesley, second edition, 1990.

5. M. Gleicher and A. Witkin. Through-the-lens camera control. In E.E. Catmull, editor, *Computer Graphics (SIGGRAPH '92 Proceedings)*, volume 26, pages 331–340, 1992.

6. L. He, M.F. Cohen, and D.H. Salesin. The virtual cinematographer: A paradigm for automatic real-time camera control and directing. In H. Rushmeier, editor, *Computer Graphics (SIGGRAPH '96 Proceedings)*, pages 217–224, 1996.

7. J.D. Mackinlay, S.K. Card, and G.G. Robertson. Rapid controlled movement through a virtual 3D workspace. In Forest Baskett, editor, *Computer Graphics (SIGGRAPH '90 Proceedings)*, pages 171–176, 1990.

8. M. McKenna. Interactive viewpoint control and three-dimensional operations. In D. Zeltzer, editor, *Computer Graphics (1992 Symposium on Interactive 3D Graphics)*, pages 53–56, 1992.

9. J.D. Mulder and J.J. van Wijk. 3D computational steering with parametrized geometric objects. In G.M. Nielson and D. Silver, editors, *Visualization '95 (Proceedings of the 1995 Visualization Conference)*, pages 304–311, 1995.

10. J.D. Mulder and J.J. van Wijk. Logging in a computational steering environment. In R. Scateni, J. van Wijk, and P Zanarini, editors, *Visualization in Scientific Computing '95, Proceedings of the sixth Eurographics Workshop*, pages 118–125, 1995.

11. P. Palamidese. A camera motion metaphor based on film grammar. *Journal of Visualization and Computer Animation*, 7(2):61–78, 1996.

12. C.B. Phillips, N.I. Badler, and J. Granieri. Automatic viewing control for 3D direct manipulation. In D. Zeltzer, editor, *Computer Graphics (1992 Symposium on Interactive 3D Graphics)*, pages 71–74, 1992.

13. K. Trovato. Autonomous vehicle maneuvering. In *Proceedings SPIE Volume 1613*, pages 68–79, November 1991.

14. R. Turner, F. Balaguer, E. Gobetti, and D. Thalmann. Physically-based interactive camera motion control using 3D input devices. In N. M. Patrikalakis, editor, *Scientific Visualization of Physical Phenomena (Proceedings of CG International '91)*, pages 135–145, 1991.

15. J.J. van Wijk and R. van Liere. An environment for computational steering. In G.M. Nielson, H. Müller, and H. Hagen, editors, *Scientific Visualization: Overviews, Methodologies, and Techniques*, pages 89–110. Computer Society Press, 1997.

16. C. Ware and S. Osborne. Exploration and virtual camera control in virtual three dimensional environments. In R. Riesenfeld and C. Sequin, editors, *Computer Graphics (1990 Symposium on Interactive 3D Graphics)*, pages 175–183, 1990.

Editors' Note: see Appendix, p. 187 for colored figures of this paper

Appendix: Colour Figures

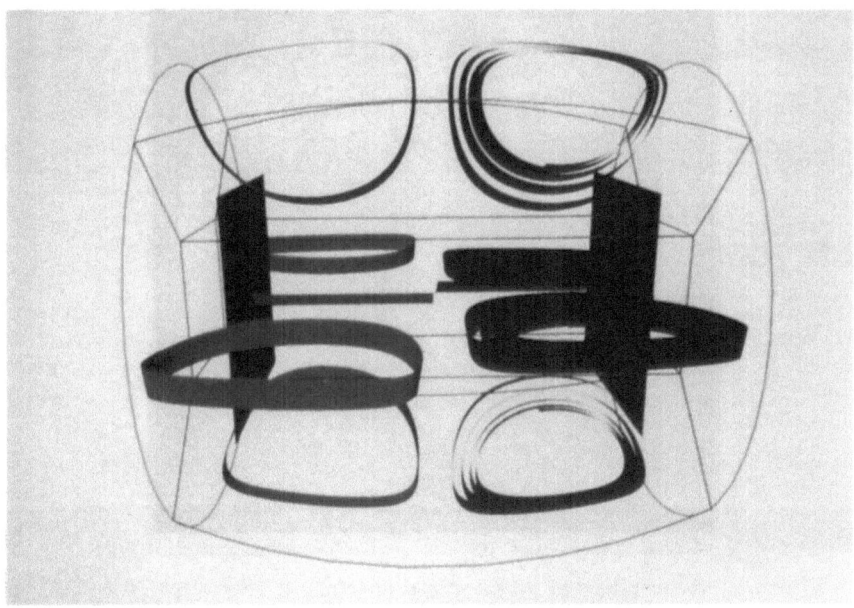

Flow in a floating-zone furnace for crystal growing. On the left hand side the stream bands are computed by the linear implicit Euler scheme and on the right hand side the classical Runge-Kutta method of order four is used (Teitzel et al., Fig. 1)

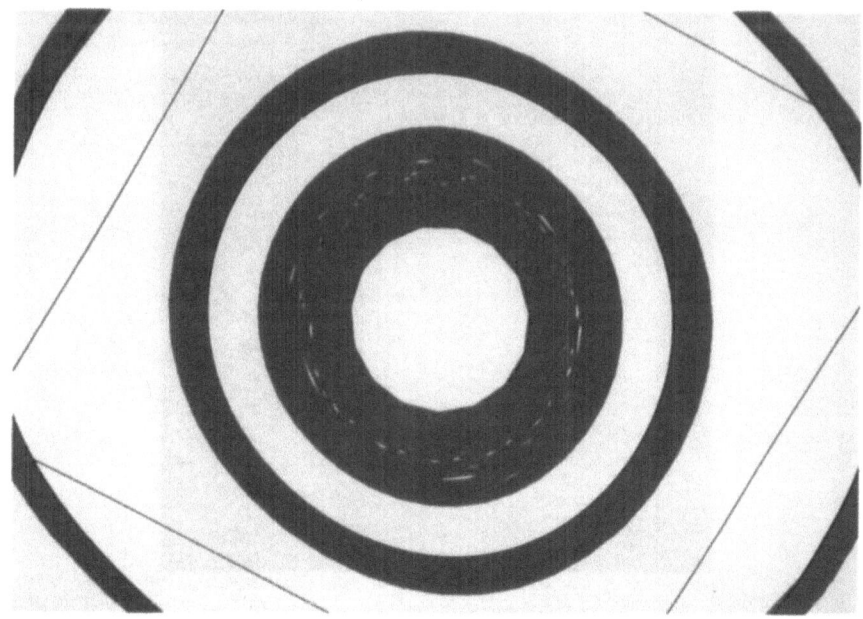

Test data set of a cylindrical air flow caused by different temperatures on the walls of the box. Here we have used both explicit and linear implicit RK4 (3). The trajectories of the implicit Runge-Kutta scheme remain closed (black circles), whereas the traces of the explicit method become thick (grey circles) (Teitzel et al., Fig. 2)

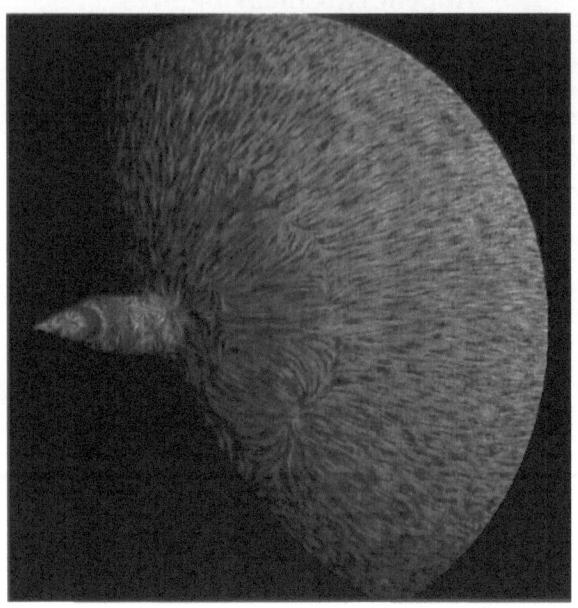

Vector projected in the normal direction of the surface
(hue: direction, saturation: magnitude) (Mao et al., Fig. 9)

Close-up view of the flow near the base of the spike
(color: distribution of density) (Mao et al., Fig. 10)

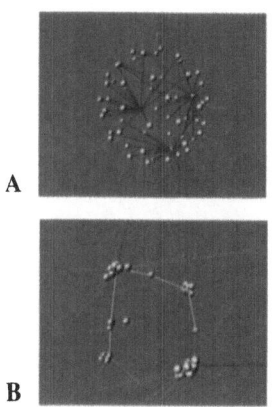

A

B

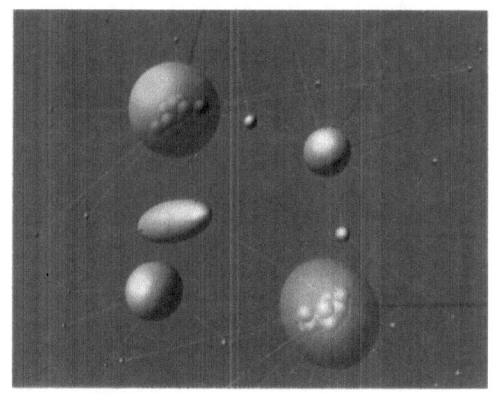

Illustration of the clustering method: Initialization (Sprenger et al., Fig. A)
Model after relaxation and highlighted minimal path between two objects (Sprenger et al., Fig. B)
Disjoined clusters as transparent ellipsoids (Sprenger et al., Fig. C)

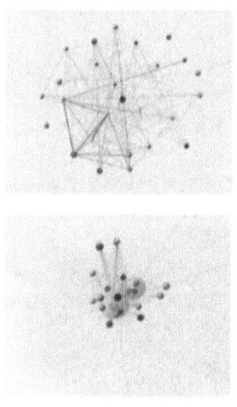

E

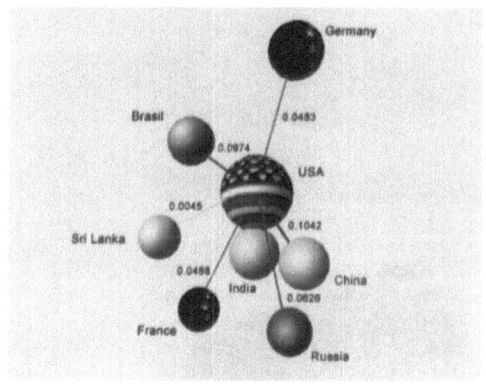

F

Visualization of agricultural productivity: Initialization (Sprenger et al., Fig. D)
Clustered energy minimum (Parameters: $R = 600$, $l_0 = 100$, $f_c = 0.02$) (Sprenger et al., Fig. E)
Discovering competitors of the United States on the world market (Sprenger et al., Fig. F)

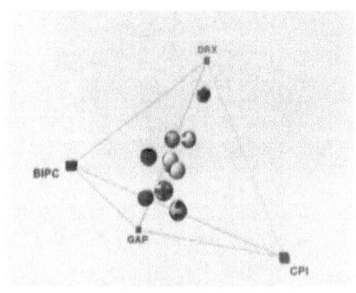

G

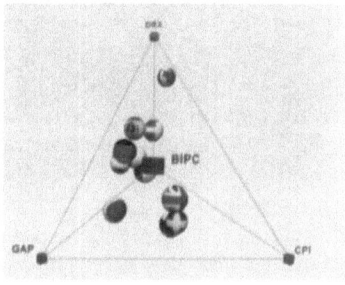

H

Influence of economic indicators onto the long term interest rates of different countries
(Sprenger et al., Fig. G)
Alternative view (Sprenger et al., Fig. H)

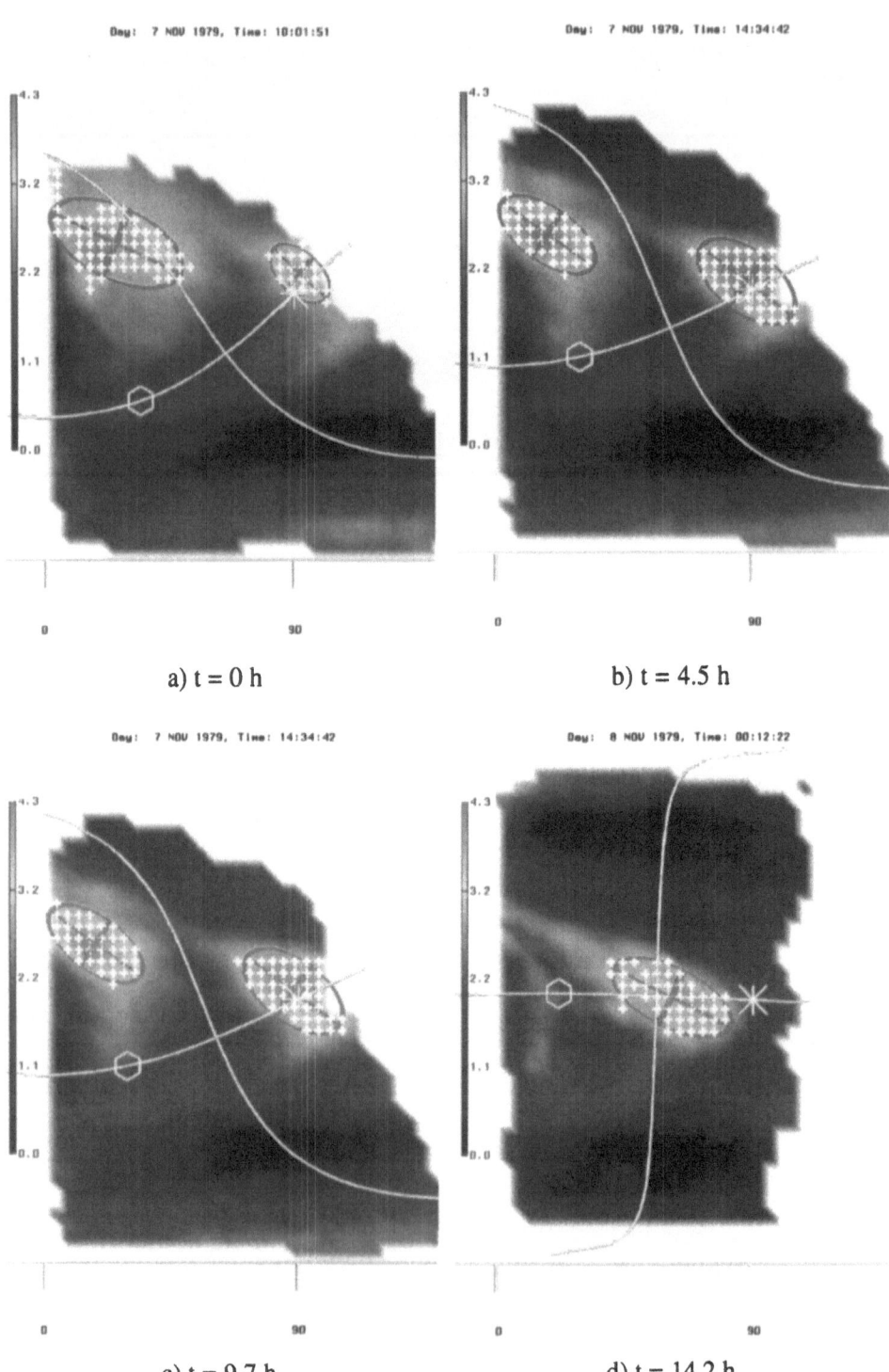

Day: 7 NOV 1979, Time: 10:01:51

Day: 7 NOV 1979, Time: 14:34:42

a) t = 0 h

b) t = 4.5 h

Day: 7 NOV 1979, Time: 14:34:42

Day: 8 NOV 1979, Time: 00:12:22

c) t = 9.7 h

d) t = 14.2 h

Determination of circulation velocity by means of feature tracking, day 507–510, |P| data, $\lambda = 365$ nm (Reinders et al., Fig. 7)

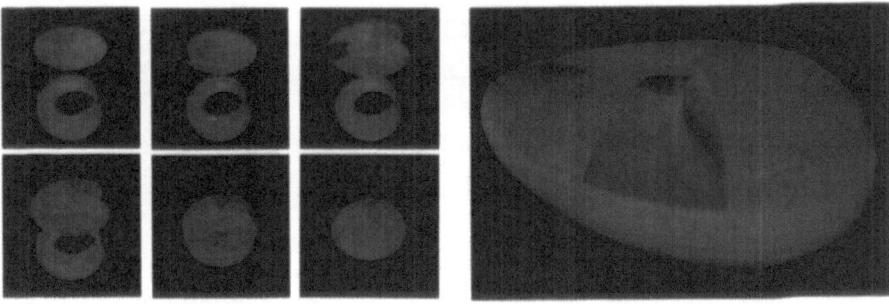

Patching Möbius band to the disc
(Klimenko, Fig. 1)

Cross cap (Klimenko et al., Fig. 2)

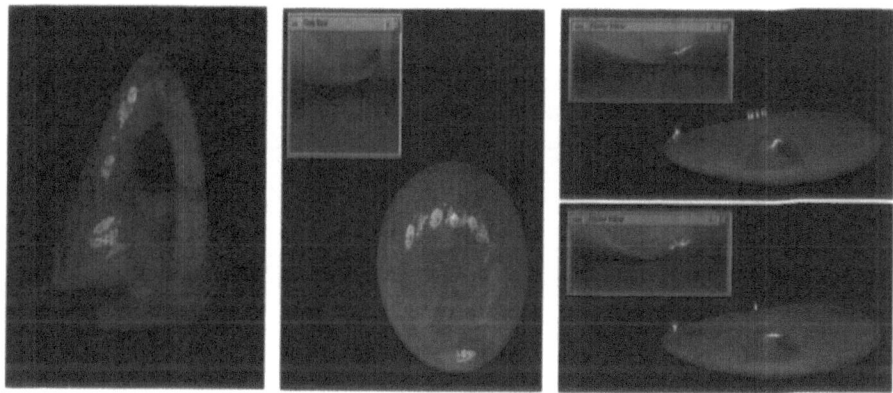

Klein bottle (left) and cross cap (right) are not
orientable (Klimenko et al., Fig. 3)

When we approach the singular
points along different directions,
we achieve different limits for
normal vectors (Klimenko et al.,
Fig. 4)

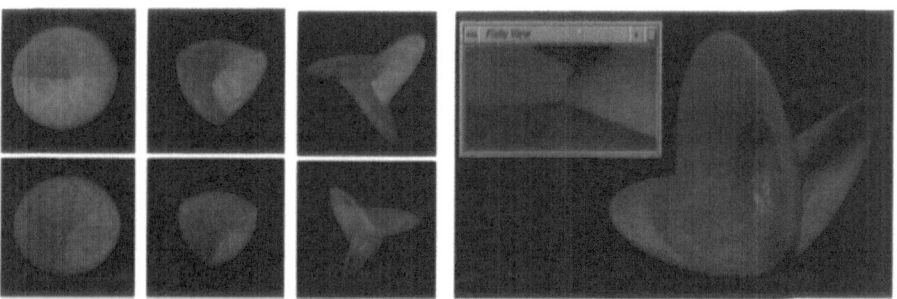

Deformation of cross cap to Boy's surface (Klimenko et al., Fig. 5)

182

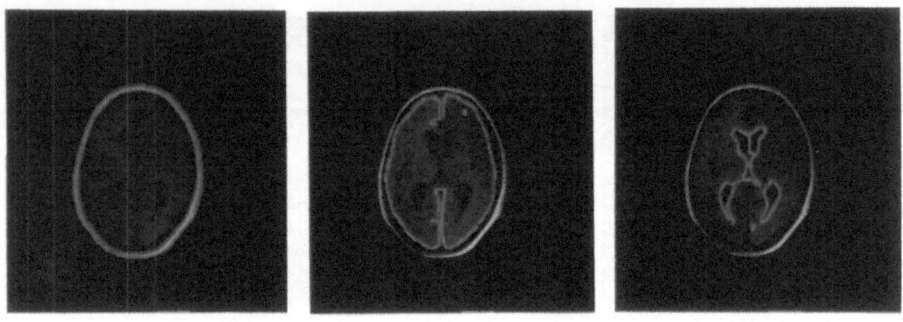

Skin (left), brain surface (middle), ventriculars (right) (Lürig et al., Fig. 1)

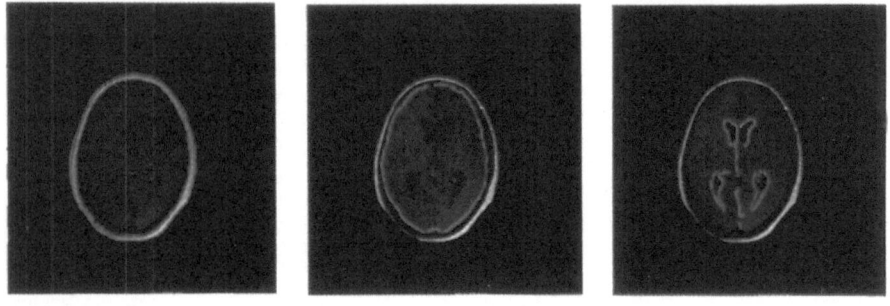

Skin (left), brain surface (middle), ventriculars (right) (Lürig et al., Fig. 2)

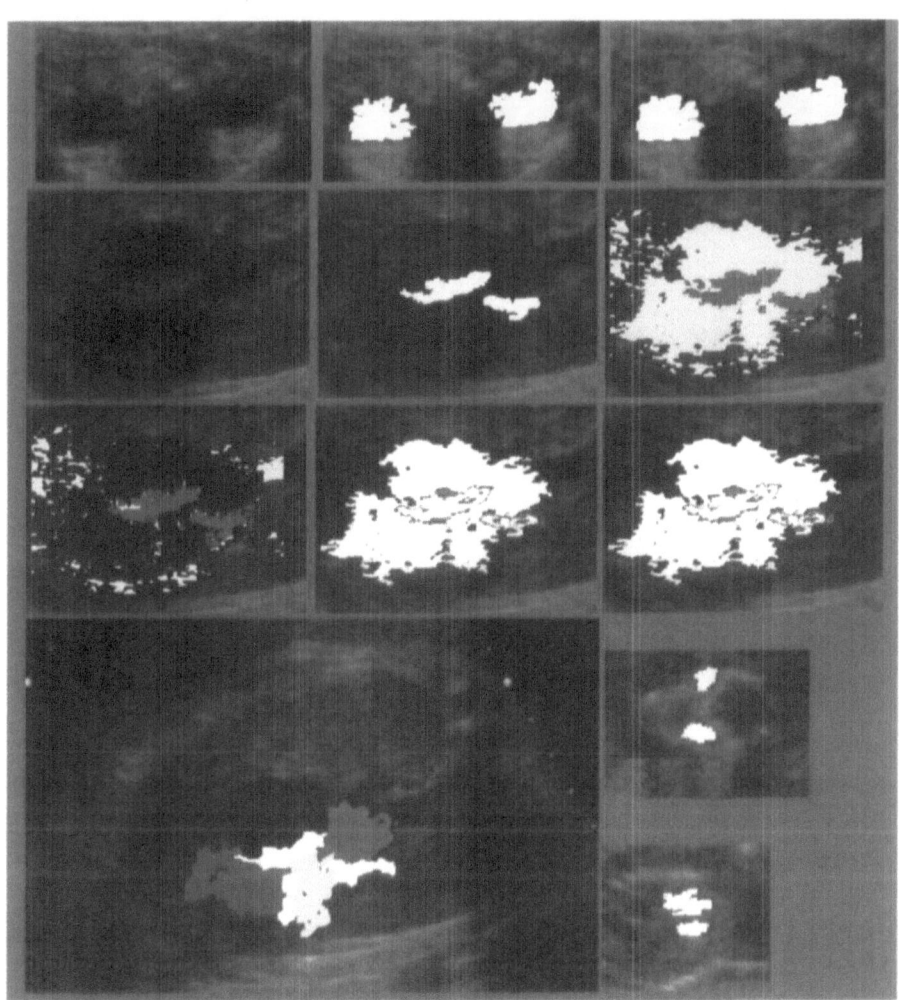

(Subramanian et al., Fig. 2)

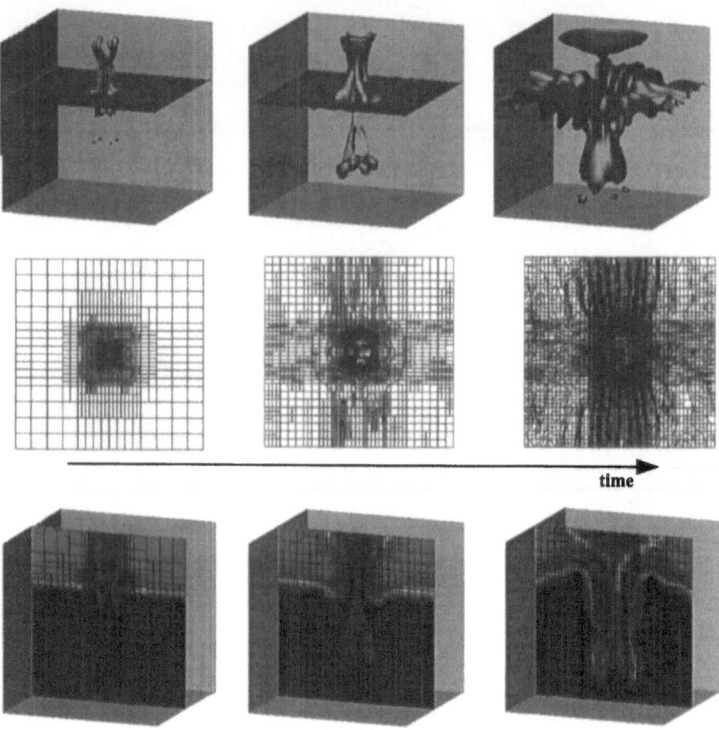

At different times porous media data is visualized using isosurfaces (above) and color shading on slices (below). Additional black lines mark intersections with element faces in a projective view (in the middle) corresponding to the isosurfaces and on the intersection plane itself (below) (Neubauer et al., Fig. 6)

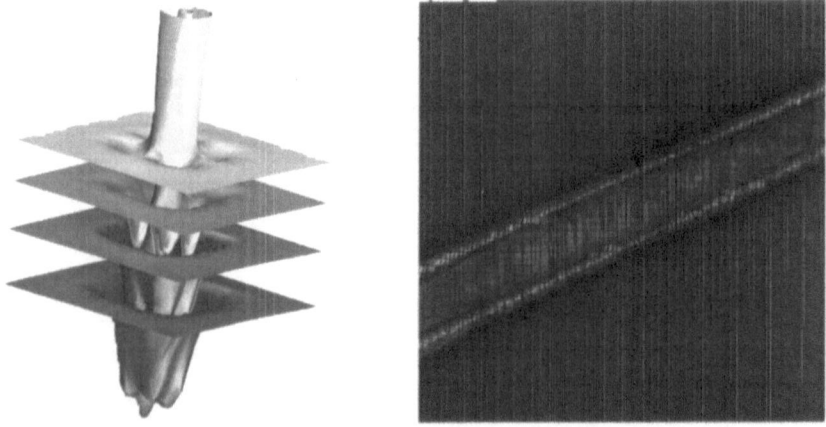

Several adaptive isosurfaces from a porous media data set and an adaptive slicing of a non-conforming hexahedral grid (Neubauer et al., Fig. 8)

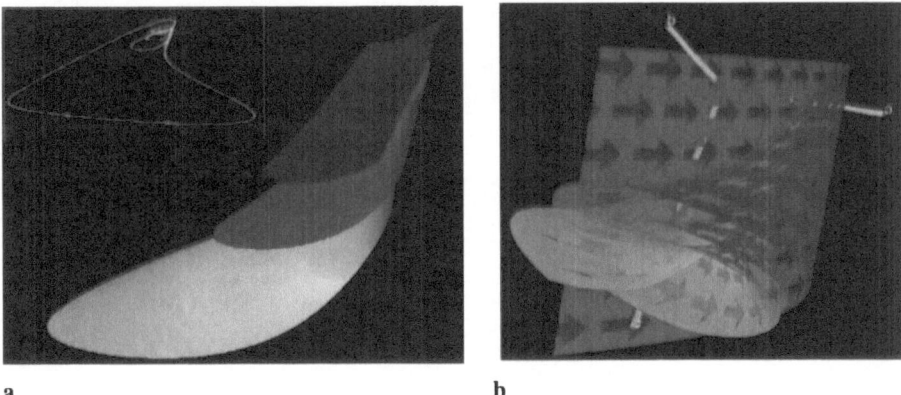

Streamline, streamsurface (**a**), and streamarrows (**b**) for mixed-mode oscillations (Löffelmann et al., Fig. 6)

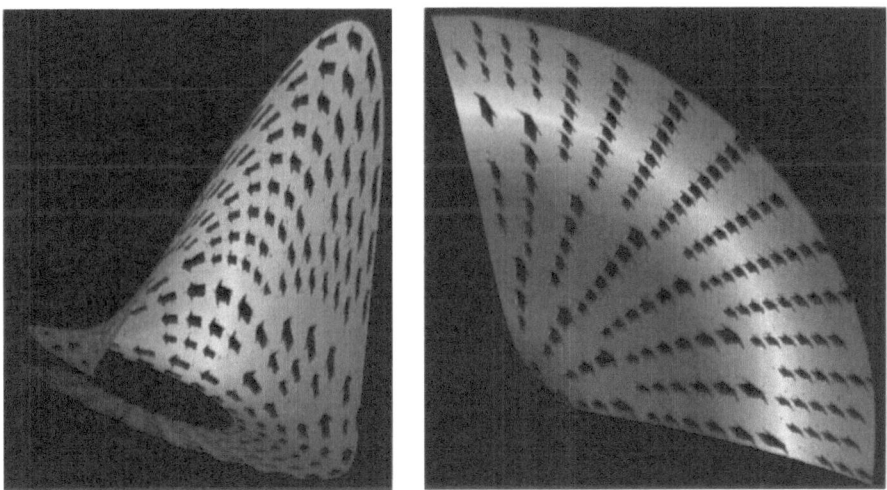

Hierarchical streamarrows, two examples (Löffelmann et al., Fig. 7)

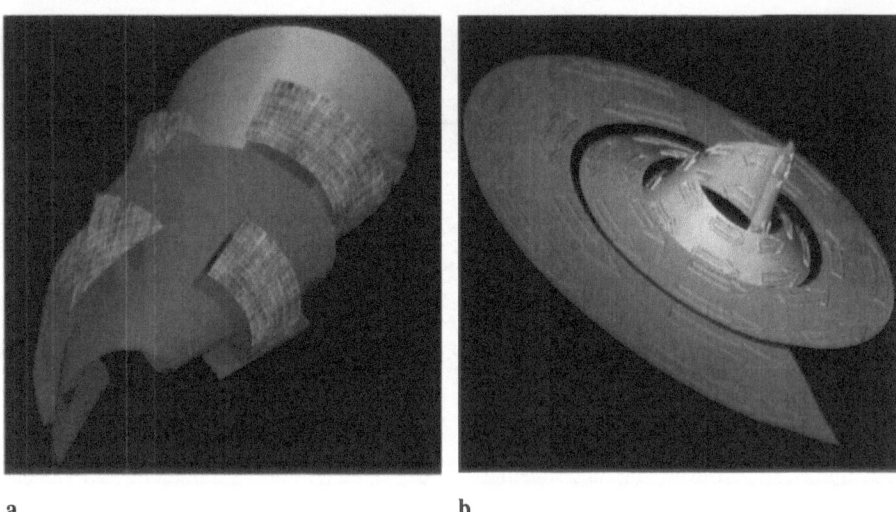

a   b

Streamarrows shifted out of the streamsurface plus anisotropic spot-noise (**a**). Streamarrows represented as 3D tubes plus color coding of velocity (**b**) (Löffelmann et al., Fig. 8)

Streamarrows similar to Shaw's hand-drawings (Löffelmann et al., Fig. 9)

Robot arm with camera objects (Mulder and van Wijk, Fig. 4)

View from the cameras present in the robot arm interface. Left the view from the camera mounted on the end effector and right the view from the camera that traces the end effector (Mulder and van Wijk, Fig. 5)

Interface to a path planning application. On the left an overview of the interface, on the right the scene from the driver's perspective (Mulder and van Wijk, Fig. 6)

SpringerEurographics

Daniel Thalmann, Michiel van de Panne (eds.)
Computer Animation and Simulation '97

Proceedings of the Eurographics Workshop in Budapest, Hungary, September 1–2, 1997
1997. 121 partly coloured figures. VIII, 203 pages.
Soft cover DM 89,–, öS 625,–
ISBN 3-211-83048-0

The contributions to this book address the problem of synthesizing the realistic movement and behaviour of human-like characters, simulated animals, fluids, and other dynamic phenomena. The animation techniques are driven by the goals of efficiency, as required by real-time interactive animations, and quality, as demanded by animations used in feature films. This series of workshops provides a high-quality international forum for the exchange of new ideas related to the themes of character animation, simulation of dynamic natural phenomena, motion capture and analysis, physically-based modeling, behavioral animation, and visualization.

Julie Dorsey, Philipp Slusallek (eds.)
Rendering Techniques '97

Proceedings of the Eurographics Workshop in St. Etienne, France, June 16–18, 1997
1997. 172 partly coloured figures. IX, 342 pages.
Soft cover DM 118,–, öS 826,–
ISBN 3-211-83001-4

The papers in this volume present new research results in the areas of finite-element and Monte-Carlo illumination algorithms, image-based rendering, ray tracing, clustering techniques, texture generation and sampling, and efficient hardware rendering. While some contributions report results from more efficient or elegant algorithms, others pursue new and experimental approaches to find better solutions to the open problems in rendering.

SpringerWienNewYork

Sachsenplatz 4-6, P.O.Box 89, A-1201 Wien, Fax +43-1-330 24 26, e-mail: order@springer.at, Internet: http://www.springer.at
New York, NY 10010, 175 Fifth Avenue • Heidelberger Platz 3, D-14197 Berlin Tokyo 113, 3-13 • Hongo 3-chome, Bunkyo-ku

SpringerEurographics

Francois Bodart, Jean Vanderdonckt (eds.)
Design, Specification and Verification of Interactive Systems '96
Proceedings of the Eurographics Workshop in Namur, Belgium, June 5–7, 1996
1996. 114 figures. XI, 383 pages.
Soft cover DM 118,–, öS 826.–
ISBN 3-211-82900-8

Making systems easier to use implies an ever increasing complexity in managing communication between users and applications. Indeed an increasing part of the application code is devoted to the user interface portion. In order to manage this complexity, it is important to have tools, notations, and methodologies which support the designer's work during the refinement process from specification to implementation. Selected revised papers from the Eurographics workshop in Namur review the state of the art in this area, comparing the different existing approaches to this field in order to identify the principle requirements and the most suitable notations, and indicate the meaningful results which can be obtained from them.

Xavier Pueyo, Peter Schröder (eds.)
Rendering Techniques '96
Proceedings of the Eurographics Workshop in Porto, Portugal, June 17–19, 1996
1996. 197 partly coloured figures. IX, 294 pages.
Soft cover DM 118,–, öS 826.–
ISBN 3-211-82883-4

Ronan Boulic, Gerard Hégron (eds.)
Computer Animation and Simulation '96
Proceedings of the Eurographics Workshop in Poitiers, France, August 31–September 1, 1996
1996. 152 partly coloured figures. X, 225 pages.
Soft cover DM 89,–, öS 625.–
ISBN 3-211-82885-0

 SpringerWienNewYork

Sachsenplatz 4-6, P.O.Box 89, A-1201 Wien, Fax +43-1-330 24 26, e-mail: order@springer.at, Internet: http://www.springer.at
New York, NY 10010, 175 Fifth Avenue • Heidelberger Platz 3, D 14197 Berlin Tokyo 113, 3 13 • Hongo 3 chome, Bunkyo-ku

SpringerEurographics

Martin Göbel, Jacques David, Pavel Slavik, Jarke J. van Wijk (eds.)

Virtual Environments and Scientific Visualization '96

Proceedings of the Eurographics Workshops in Monte Carlo, Monaco, February 19–20, 1996, and in Prague, Czech Republic, April 23–25, 1996

1996. 169 partly coloured figures. VIII, 324 pages.

Soft cover DM 118,–, öS 826,–

ISBN 3-211-82886-9

Bodo Urban (ed.)

Multimedia '96

Proceedings of the Eurographics Workshop in Rostock, Federal Republic of Germany, May 28–30, 1996

1996. 71 figures. VII, 178 pages.

Soft cover DM 85,–, öS 595,–

ISBN 3-211-82876-1

Remco C. Veltkamp, Edwin H. Blake (eds.)

Programming Paradigms in Graphics '95

Proceedings of the Eurographics Workshop in Maastricht, The Netherlands, September 2–3, 1995

1995. 41 partly coloured figures. VIII, 172 pages.

Soft cover DM 94,–, öS 655,–

ISBN 3-211-82788-9

Philippe Palanque, Rémi Bastide (eds.)

Design, Specification and Verification of Interactive Systems '95

Proceedings of the Eurographics Workshop in Toulouse, France, June 7–9, 1995

1995. 153 figures. X, 370 pages.

Soft cover DM 118,–, öS 826,–

ISBN 3-211-82739-0

 SpringerWienNewYork

Sachsenplatz 4-6, P.O.Box 89, A-1201 Wien, Fax +43-1-330 24 26, e-mail: order@springer.at, Internet: http://www.springer.at
New York, NY 10010, 175 Fifth Avenue • Heidelberger Platz 3, D-14197 Berlin Tokyo 113, 3-13 • Hongo 3-chome, Bunkyo-ku

SpringerEurographics

Martin Göbel (ed.)
Virtual Environments '95

Selected papers of the Eurographics Workshops in Barcelona, Spain, 1993,
and Monte Carlo, Monaco, 1995
1995. 134 partly coloured figures. VII, 307 pages.
Soft cover DM 119,–, öS 832,–
ISBN 3-211-82737-4

Demetri Terzopoulos, Daniel Thalmann (eds.)
Computer Animation and Simulation '95

Proceedings of the Eurographics Workshop in Maastricht, The Netherlands, September 2–3, 1995
1995. 156 partly coloured figures. VIII, 235 pages.
Soft cover DM 98,–, öS 688,–
ISBN 3-211-82738-2

Riccardo Scateni, Jarke J. van Wijk, Pietro Zanarini (eds.)
Visualization in Scientific Computing '95

Proceedings of the Eurographics Workshop in Chia, Italy, May 3–5, 1995
1995. 110 partly coloured figures. VII, 161 pages.
Soft cover DM 94,–, öS 655,–
ISBN 3-211-82729-3

Patrick M. Hanrahan, Werner Purgathofer (eds.)
Rendering Techniques '95

Proceedings of the Eurographics Workshop in Dublin, Ireland, June 12–14, 1995
1995. 198 partly coloured figures. XI, 372 pages.
Soft cover DM 118,–, öS 826,–
ISBN 3-211-82733-1

Martin Göbel, Heinrich Müller, Bodo Urban (eds.)
Visualization in Scientific Computing

1995. 150 figures. VIII, 238 pages.
Soft cover DM 118,–, öS 826,–
ISBN 3-211-82633-5

SpringerWienNewYork

Sachsenplatz 4 6, P.O.Box 89, A 1201 Wien, Fax +43 1 330 24 26, e mail: order@springer.at, Internet: http://www.springer.at
New York, NY 10010, 175 Fifth Avenue • Heidelberger Platz 3, D-14197 Berlin Tokyo 113, 3-13 • Hongo 3-chome, Bunkyo-ku